FISH TALES

The Mystery of the Snapper Bump
and other stories from
Australian waters

For: Daniel Zivilli

A small tome to add
to your fine collection.
Best wishes

Julian Pepperell

FISH TALES

The Mystery of the Snapper Bump and other stories from Australian waters

Julian Pepperell

RANDOM HOUSE AUSTRALIA

Published in Australia by
Random House Australia Pty Ltd
20 Alfred Street, Milsons Point, NSW Australia, 2061
tel 61 2 9954 9966
fax 61 2 9955 3381
http:/www.randomhouse.com.au

Sydney New York Toronto
London Auckland Johannesburg
and agencies throughout the world

First published in 2001

Pepperell, Julian G.
Fish tales.

Bibliography.
Includes index.
ISBN 1 74051 433 5.

1. Fishing - Australia. 2. Fishing. I. Title.

799.1

Publisher: James Mills-Hicks
Design: Lena Lowe and Sue Rawkins
Editor: Lucy Tumanow-West
Cover: Nicole Struik
Production: Linda Watchorn and Angela Alegounarias
Publishing Assistant/Index: Anabel Pandiella

Printed by Sing Cheong Printing Co. Ltd, Hong Kong
Film separation: Pica Digital Pte Ltd, Singapore

Cover photograph: Catfish *(Rudie Kuiter)*
Half title: The redcoat squirrelfish *(Rudie Kuiter)*
Title page: Coral Trout *(Bernard Yau)*
Opposite: Soldier-fishes *(Rudie Kuiter)*
Page 10: Dwarf Hawk Fish *(Kelvin Aitken)*
Page 52: Soft Coral *(Kelvin Aitken)*
Page 76: Black Marlin *(Peter Bristow)*
Page 100: Lemon Damsel *(Kelvin Aitken)*
Page 154: Great White *(Kelvin Aitken)*
Page 172: Big Eye Trevally *(Kelvin Aitken)*

To Daniel, who surfs with dolphins.

Contents

Introduction 8

CHAPTER 1 FISH TALES

Revenge of the Dolphin Fish 12
Mushy Sushi Anyone? 15
Old Man Snapper? 18
The Mystery of the Snapper Bump 21
The Big Sting 24
Death, Where is thy Sting? 27
Poison Poisson 30
A Long Way from Home 33
Pink Bolt from the Blue 36
Colossal Calamari 39
And Thereby Hangs a Tail 42
Blast from the Past 45
Voyages of Discovery 48

CHAPTER 2 BIODIVERSITY AND FISH

The Diversity of Life ... and Fish 54
What's in a Name? 57
Cosmopolitan Fish 60
Warm Blooded Fish 63
Odds and Cods 66
Dragons, Ghosts and Horses 69
Pickled Fish 73

CHAPTER 3 BIG FISH DOWN UNDER
Of Baby Bills 78
Do Marlin Change Sex? 81
Mega Yellowfin 84
Tiny Tuna 87
All in the Genes 90
Sail Away 93
Twice Bitten, not Shy 97

CHAPTER 4 FASCINATING FISH FACTS

And Fish Might Fly 102
Smell and Taste in Fish 106
Built for Speed 109
Mucus and Slime 112
The Colour Purple 115
Stripe Me Lucky 118
The Oldest Fish? 121
Telling the Boys from the Girls 124
Like a Fish out of Water 127
Raking It In 130
The Amazing Argonauts 133
Sticking to the Job 136
The Empty Nest 139
School's In 142
Fronts and Slicks 146
To Sleep, Perchance to Dream 149

CHAPTER 5 SHARK STORIES

Shark Attack 154
Sizes of Sharks: Fact versus Fiction 157
The Weird, Wonderful Hammerheads 160
The Cookie Monster 163
A Tale of a Whale (shark) 166
The Mysterious Megamouth 169

CHAPTER 6 CONSERVATION AND FISHING

Do Fish Survive Catch and Release? 174
Survival of the Fittest 177
What is a Normal Fishing Year? 180
Fish Sauce 183
Flotsam and Jetsam 186
Is Fishing OK? 189
Once Bitten, Twice Shy? 192
Thoughts on Shark Finning 195

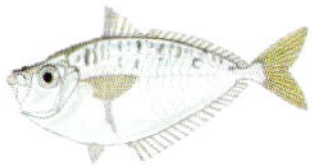

Index 198
Acknowledgements 207

Introduction

The study of the marine world could easily consume thousands of lifetimes. So far, mine has been just one such fortunate life spent in pursuit of an understanding of life under the seas and oceans. My particular fascination has been with the fishes, especially the large, surface-dwelling fishes, marlins, tunas and sharks, which have always held a special allure. But the whole gamut of marine life has nevertheless held me spellbound since my earliest recollections.

I grew up on the foreshores of Port Phillip Bay, Victoria, poking around in rock pools for hours on end; observing the little blennies, starfish, anemones and octopuses with intense curiosity. Fishing from the local jetties also became a favourite pastime, thanks to expert tuition from my father and grandfather. Another early memory is of avidly browsing the pages of the angler's bible of that time, Fish and Fisheries of Australia by Theo Roughley. I especially liked the author's handwritten inscription to my father: "May the fish that get away lure you back again."

It was therefore appropriate, I suppose, that years later I would embark on a career in the study of fish, joining New South Wales Fisheries with the title of Marine Angling Biologist, the first such position in Australia.

For the next decade and a half, interspersed with field trips, I was invited on numerous occasions to give talks (I refrain from using the term "lectures") to fishing clubs and other gatherings of the angling community. The topics were wide-ranging, usually covering aspects of the biology of fish species that were of interest to the particular group being addressed.

Most of the time, the audiences were attentive and polite, even interested, dare I say. However, what I invariably noticed was that at the end of the talk, when the club president would say "Any questions?", everyone would come to life, even those at the back of the room who had been quietly nodding off.

Yes, the "Any questions" cue was the one they had been waiting for. Sure, the talk may have contained some interesting snippets, but here was a captive marine biologist offering to answer the question that had been gnawing away at the back of their mind for years. Why? What? How? When? It became obvious very quickly that anglers are extremely curious about the fish they catch, or at least try to catch.

After this realisation, I learned to shorten the actual talks and allow plenty of time for the incredible range of questions that were invariably asked (with the hope of finishing the meeting before midnight).

When you think about it, anglers spend a great deal of time pondering the world of their finny adversaries. Isaac Walton called fishing the contemplative man's sport, and there is no denying that it gives one plenty of time to contemplate: the vagaries of angling success being well nigh imponderable, when the fish aren't biting, the mind invariably wanders to the watery world below. Pet theories take shape over time and, eventually, those theories might at last be placed under the nose of the guest boffin giving the talk at the local club. And so the notion for this book was born.

As I was putting together the material for this book, it became clear to me that a vast amount of fascinating information has been collected by armies of dedicated scientists and dutifully published in the scientific literature. Unfortunately, though, once published in that format, the dry, scientific papers all too often gather dust on the library shelves of academia, being lost to the general public to all intents and purposes. Certainly, fisheries management decisions are often made on the basis of such research, but the basic results, which are often of great interest to the lay person, never make it into the general media. I hope this book will help to fill that gap.

Many of the essays presented are the results of questions asked of me during public talks, and also through writings and feedback in popular print and electronic media. Letters, emails and phone calls have come from people of all age groups, from school children through to retirees, and from all walks of life. Questions have come not only from those who love the joys of recreational fishing, but also divers, beachcombers and nature lovers, all of whom have their own fascination with the wonderful marine world.

As you will discover, no-one can answer some questions with absolute certainty, simply because the research has not yet been carried out. After all, life in the seas and oceans is hugely complex, and not easy to study. Nevertheless, researching the many and varied topics that make up the contents of this book has always been fascinating, not to mention fun.

The questions and topics covered are as varied as the subjects themselves. Do flying fish flap their wings? Why do many fish have stripes? What is the largest shark? What makes the crater-shaped wounds found on many large fishes? How fast do fish swim? What animal produces the paper nautilus shell? Do fish sleep? How many eggs does a marlin produce? Do fish care for their young? And so on.

What does become clear is that the more you fine out about this field, the more questions there are to be asked.

I hope the short pieces presented here will help to entertain, enlighten and educate – and in doing so, stimulate further curiosity and inquiry. If so, then this book will have achieved its end.

Julian Pepperell
April 2001

Fish Tales

It's surprising what an important part fish play in the lives of many people. Anglers, divers, students, aquarists and nature lovers alike, all have a certain fascination about the finny denizens of the watery depths. In the course of studying, writing and lecturing about fish, I've been asked all kinds of questions about fish biology. Here are some of the musings which have resulted.

The wonders of marine life never cease to amaze

Revenge of the Dolphin Fish

It's funny. Science in its essence is an exacting discipline whereby one sets out to try to disprove a particular hypothesis by means of rigorously designed experiments. But Murphy dictates that discoveries about the world around us don't always happen that way. Often, good old-fashioned luck plays a role too.

Take for instance this series of events which led to some fascinating observations and musings about the pelagic food chain and the distribution and abundance of two species of fish.

It began on a day spent out off Port Stephens during an interclub game fishing tournament. Billfish were the target for the day, but as luck would have it, nary a beak was seen, at least, not outside the boat. In fact, we only had two strikes for the day, and both turned out to be very nice dolphin fish, or mahi mahi if you prefer. The first fish was about 10 kg, and the second, over 20 kg. Dolphin fish are excellent table fare, so this was not bad compensation for bombing out on billfish.

I am naturally always very interested in the diets of any fish I catch, and dolphin fish in particular often have unusual food items in their stomachs. So, opening the first fish, I noted the contents: several small toadfish, three squid beaks, a couple of tiny scads, some plastic, and, lo and behold, a perfect specimen of a baby broadbill swordfish about 12 cm long. Even though it didn't look like an adult swordfish, I knew the unmistakable characteristics of a long slender upper and lower jaw, still lined with teeth, and body covered with rows of protective denticles. The swordfish must have only just been eaten because it was unmarked, and what an amazing sight it was. Its body had several broad, vertical bars, and its gill plates were iridescent blue and purple; a far cry from the rather dull colours of the adult.

Stomach contents of a 10 kg dolphin fish Coryphaena hippurus. *Note the piece of plastic, and especially the beautiful baby broadbill swordfish.*

When the second, larger dolphin fish was caught later in the day, it was quickly hoisted up onto the operating table, and this time I hit the jackpot. Three juvenile broadbill, ranging from 20 to 45 cm, filled the stomach, causing much scratching of chins and clicking of shutters. Now under

normal circumstances, one might expect dolphin fish to be on the menu of the billfishes, but here we were witnessing the reverse situation – the piscatorial equivalent of "man bites dog".

A small broadbill swordfish, Xiphias gladius. *Normally a feared predator at the top of the food chain, but when small, just another morsel for other fishes.*

I subsequently took the opportunity to look at other dolphin fish caught during the tournament, and can report that of eight fish examined, seven had eaten juvenile broadbill. Interestingly, they usually contained more than one, the most being six little swordies in one stomach.

What are we to make of these fortuitous observations? Well, to begin with, that particular season had been a very good one for large dolphin fish off eastern Australia, with large numbers of double figure fish (in kilos that is) being caught from Cairns to Bermagui over late summer/early autumn. With so many feeding on baby broadbill, it is reasonable to conclude that these too were very common at the time, and presumably were spawned not too far away within the last few months. But are there always concentrations of juvenile swordfish off New South Wales at this time of the year that would normally go unnoticed in the absence of marauding dolphin fish? Those sorts of questions are difficult to answer, but they set the scene for some interesting research in future years.

Juvenile billfish are not encountered very often, but when they are, a lot of information can be gleaned from them. For example, studies on the age and growth of fish (billfish included) are based on examining hard parts such as otoliths (ear bones) for banding patterns, either annually, or even daily. Very small billfish (less than say 5 kg) are needed to help in validating the periodicity of such bands, since they would almost certainly be less than one year old. Also, information on times of spawning, nursery areas and early dispersal rates can be determined by analysing occurrence patterns of baby billfish.

Probably one of the reasons that juvenile billfish are so rarely seen is that they grow so fast and are therefore not small for very long. Another possibility is that they are (fortunately for them) not prone to being caught in or by normal fishing or sampling gear. Enter the dolphin fish and other pelagic predators that are very efficient

"samplers" of the upper layers of the ocean, and in doing so, can provide us with some rare glimpses into the life cycles of their larger cousins; if we know where to look.

I mentioned earlier that one of the dolphin fish I examined had a piece of plastic in its stomach. A study of the diet of dolphin fish from the Atlantic Ocean in 1984 examined the stomachs of over 2600 fish and found, not surprisingly, a wide variety of fish, crustaceans and squid in most stomachs. However, a surprising finding was that a significant number of stomachs contained plastic or other man-made items such as tar balls, cigarette filters, nylon string and even a light bulb. It is generally thought that plastic and fish don't mix very well, and ingesting plastic might well cause the demise of the fish. However, the authors of the above study made the rather novel proposal that, since dolphin fish eat such a wide range of non-food materials, they should be regarded as indicator species for offshore water quality, particularly in areas threatened by oil spills or subject to ocean dumping. Perhaps not such a silly idea.

Above: An adult dolphin fish, or mahi mahi. One of the fastest growing fishes in the ocean, dolphin fish are voracious eaters.

Right: A rare catch. A broadbill swordfish about to be released.

Mushy Sushi Anyone?

Can you imagine a more unpleasant experience than catching a nice fish, icing it down to prepare it for a culinary masterpiece, and then putting the filleting knife into the fish to reveal a pasty, gooey mass where the flesh should be. This is exactly what happened to someone I heard of who had caught a lovely 7.5 kg dolphin fish off Cairns.

A *nice dolphin fish, but will it be a culinary disaster?*

I have been interested in this phenomenon for some time, especially during work on yellowtail kingfish, which also show this alarming condition at times. In the case of kingfish, however, the mushiness is not apparent in raw fish, but only occurs when the fish is cooked (even more annoying than finding it in raw fish). The other interesting thing about the condition in kingfish off the east coast of Australia is that it only occurs in fish caught from about Coffs Harbour north. Kingfish caught off Brisbane are always suspect, although frustratingly, some individuals will be perfectly all right. It can be a real lottery, then, as each guest at the dinner party opens the aluminium foil to reveal either a succulent feast, or impromptu fish soup.

So, what is the cause of all the mush, I mean fuss? Parasites, but they are not the usual worm or creepy crawly normally associate with the word "parasite". In fact, they are single-celled protozoan parasites known as Myxosporideans, which infect the fish in their millions. (Protozoan means "first animal", and other protozoans that are probably better known are amoebae and the bug that causes malaria.)

In the case of kingfish, the culprit is known as *Unicapsula seriolae* (named after the kingfish, *Seriola*), and fortunately, we know a little about it, thanks to work done by Rob Lester of Queensland University. The little animal enters its host in an as yet unknown manner, travelling through the organs to eventually form reproductive spores. The spores of the organism cause the problem, and these form in lines up to 7 mm long and 0.2 mm wide lying inside muscle fibres. (The spores themselves are absolutely microscopic.) When the fish is alive, the spores don't do major damage to the muscle, but produce an active chemical known as an enzyme which causes the dissolving (histolysis) of the muscle cells when the host dies. In the case of *Unicapsula*, the enzyme apparently becomes activated when cooking raises the temperature of the fish. One interesting finding was that the effect was greatest in slow cooking, and the recommendation therefore is to microwave your kingfish in suspect areas, rather than baking or putting it on the barbecue.

Yellowtail Kingfish
Seriola lalandi
180 cm
110 kg

Temperate coastal waters of the Indo-Pacific, including California, New Zealand, South Africa and Australia. In Australia, it ranges from Rockhampton, Queensland to Geraldton, Western Australia. North of Coffs Harbour off eastern Australia, kingfish have a high incidence of infection with the protozoan parasite Unicapsula, which turns the flesh into mush when cooked.

Returning to the dolphin fish, this is the first time I have heard of this problem occurring in this species on the east coast. However, exactly the same symptoms are known from Albany, Western Australia, mainly because of an outbreak in an experimental dolphin fish farm over there some years ago. Again, the spores of a protozoan (Myxosporidean) parasite were identified as the cause, but a different one this time. The name of the offending parasite is *Kudoa thyrsites*, and its effects are so similar to the symptoms described, that I would put good money on it, or a very close relative, being the culprit in this case.

The difference between the dolphin fish parasite (*Kudoa*) and the one infecting kingfish (*Unicapsula*) is that the spores of *Kudoa* remain active for several days at 4°C, that is, while the fish is nicely iced down. It produces a similar histolytic enzyme to that of *Unicapsula*, but it is active at low temperatures, with the result that an iced fish will slowly turn to slush.

Why on earth does this happen? Well, it is possible that the protozoans start to secrete the enzymes in response to the death of their host on purpose, to break down the muscle cells that surround them. In this way, the spores would be quickly released to the outside world to begin a new generation. After all, it would not be in the parasite's interest to go down with the ship, as it were.

The complete life cycles of these organisms are not known between the death of one host and re-infection of another. Many parasites have complex life cycles that often involve the infection of an intermediate host, commonly an invertebrate such

as a snail or crustacean, before re-infecting the main host, and this could well be the case for these protozoans.

A mouth-watering display of sashimi and sushi, but sometimes, raw fish may harbour unwelcome guests.

In both the Queensland and Western Australian studies of this phenomenon, although considerable actual or potential economic harm might be caused by these parasites, no real solutions to the problem were forthcoming. Farming of dolphin fish and kingfish shows considerable promise, but if it is to become a reality in this country, then we will certainly need to find ways of controlling these unwelcome invaders. I should stress that the vast majority of fish parasites are quite specific to their host species, and so, cannot infect humans in the same way. This is especially so if the fish is cooked first (although in this case, one would not wish to eat the resulting mush). A final note on the subject, though. A small number of fish parasites are transmissable to humans through eating of raw fish. Fortunately, the incidence of such cases is extremely low, and has never dissuaded me from tucking into a delicious plate of sashimi or sushi!

Old Man Snapper?

Talking on the fishing club circuit over many years, some generic questions seem to crop up very frequently. Probably the most common of these is "How old is that fish?"

Two things affect the answer to this. Firstly, growth rates in fish do vary considerably depending on environmental conditions, and secondly, it is very difficult to age a fish without having the specimen. However, with a bit of detective work, a reasonable guess can usually be made. It's worthwhile explaining the process involved in ageing fishes generally. Several methods have been devised, but two are used most commonly.

The first involves measuring the lengths of many fish from a population, and then plotting the frequency of each size group. Because most fishes have one distinct annual spawning period, sizes of fish in a population tend to be bunched into groups, or size classes. More often than not, these size classes correspond to groups of fish of similar ages, each separated by a year, and are therefore termed year classes. This method is particularly useful during the first few years of life, when the year classes tend to be clearly distinguishable. As the ages increase, the distinctions between the year classes become less apparent.

"How old is that fish?" It would be nice to be able to answer with certainty, but many variables affect the growth rates of fish.

The second, more widely known, method of ageing fishes is by means of counting rings or bands on various bony structures, such as the scales, or better, the otoliths (tiny bones in the middle ear). Because fish are cold blooded, they grow in spurts (faster in summer and slower in winter). Their bones and scales grow in proportion, and this is reflected by distinct banding patterns within those structures. Most people are familiar with the phenomenon of counting rings on a tree to determine its age (before it was cut down), and this is also the case for fish, provided that the rings do in fact represent yearly events. This is where care is necessary, since events other than winter can cause rings to appear on scales and otoliths.

For example, spawning is a major event in the life of a fish, and metabolic changes during this time can cause a ring to be laid down in the scale or otolith. Similarly, drought, floods or other environmental changes can also cause changes in growth of the whole body, and of the bony structures. For these reasons, it is important to collect scales and or otoliths from large numbers of fish covering all sizes over as long a period as is practicable to allow for these sorts of possible effects to be taken into account.

The Australian Snapper is a relatively long-lived species with a life span of up to 30 years.

Surprisingly, very little work has ever been done on the growth rates of snapper off the Australian east coast. Fortunately however, Gary Henry of NSW State Fisheries conducted a study in the mid 1980s on the snapper population off Coffs Harbour. Gary aged over 1700 fish by counting rings on scales and otoliths, and verified his ages by examining the size structure of the population of smaller fish. Perhaps not surprisingly, he came across very few large fish, but his sample of large fish was sufficient to give an estimate of the age of a 9 kg fish at between 16 and 22 years. This may seem imprecise, but it is quite normal to be less certain about ages of fish the larger they are, since the rings cram closer together, and also since size at a given age becomes much more variable through time. The age of a fish of the same size taken in Victorian or South Australian waters would be quite different. Unfortunately, we do not as yet have reliable age estimates from those areas.

Arriving at an age for any particular fish without examining it is a guess at best, and relies heavily on previous repetitious and meticulous work having been done on that species in the fisheries laboratories around Australia. It is therefore of particular interest to note that the Australian Fish Ageing Facility within the grounds of the Marine Science Laboratories of the Victorian Fisheries Department at Queenscliff, headed by Dr David Smith, exists solely to age fish from hard parts sampled from fish populations around Australia. Priorities necessarily lean towards commercial species of fish but this facility is proving to be of immense importance to recreational fisheries

Anglers are often curious about the age of the fish they catch.

as well. The perennial question "How old is that fish?" hopefully will now be more readily answered.

Finally, if you do catch a fish and are curious about its age, and would like an expert's opinion, take down its length in addition to its weight. All fish should be measured in millimetres "on the flat", ie in a straight line from the tip of the snout or projecting lower jaw (mouth closed) to the fork in the tail. Some scales taken from behind the pectoral fin would also assist, or better still, the otoliths (ear bones, or "jewels") from inside the skull. Pop the hard parts (scales or otoliths) into an envelope, and carefully label it with the measurement details as well as the date and location of capture. Your local Fisheries Department would then be the next port of call, where hopefully, an expert technician or biologist will provide you with a good "guesstimate" of the age of your fish, especially if that particular species has been studied in any detail.

The Mystery of the Snapper Bump

Why does the Australian snapper have a bump, or hump, on its head? Or more to the point, why do some snapper have humps and others don't? For example, large snapper (10 kg plus) from the Bay of Islands in New Zealand, or from Whyalla in South Australia, are "bumpless", while similar sized fish from parts of the New South Wales and Victorian coasts possess huge humps on their foreheads.

The hump is in fact a true thickening of the bone on the forehead, while extra-large "old man" snappers also have fleshy expansions of their lips and noses, sometimes to a grotesque degree. Interestingly, it seems that the two go together, no hump, no fleshy lips and nose.

But why do the fish with the humps have them at all? Of course, just about everyone who catches snapper, commercial or recreational, has his or her own idea of the phenomenon. By far the most common theory is that the hump is used to knock shellfish off reefs. A colleague who has seen his fair share of big snapper reckons that snapper with humps are always caught over reefs, and certainly commercial fishermen have related the same thing to me, so there may be something in this observation. I also have it on good authority that relatively small snapper caught from particular reefs in northern New South Wales will have large humps, while larger fish from nearby reefs will not. The reef effect is also borne out by the fact that in New Zealand snapper (no humps) tend to be caught over shaly bottom, and I believe that the big humpless snapper of South Australia also live in non-reef areas.

Two similar sized snapper, Pagrus auratus, *one with a large bump, and one without. Why is it so?*

So what about the theory involving knocking shellfish off reefs? I'm afraid I don't subscribe to that idea, since the humps are nearly always covered with normal skin and scales, and surely this area would therefore be injured or at least show signs of scarring if the fish were head-butting rocks and sharp oysters and mussels all day long! Besides, the growth of the bone is normal to all external appearances, and not the result of the formation of calluses from constant abuse.

I have a pretty good library on fish and fishing, but scouring the shelves I couldn't find a single book that put forward a possible reason for this feature, even though many authors were certainly aware of the phenomenon.

THE PEPPERELL THEORY

It's now time to put in my two cents' worth, so here goes. John Ashley, a fishing colleague, is a snapper expert, and when he catches a few prime specimens off southern Sydney (which is pretty regularly), he and I head for a great little Thai restaurant where they do the best whole fried snapper in the world (especially if you bring your own). After every delectable morsel is picked from the bones, my inquiring mind tends to take over and I have noted that in virtually every specimen, marked swellings on certain bones are evident. The swellings are mostly observed on the neural spines, that is, the long, bony extensions of the vertebral column (backbone) which support the back muscles. Often, there is a neat, round swelling on just about every spine, looking like little balls of bone skewered by the spines. Looking at them more closely, the swellings are obviously part of the bone itself, and have apparently formed as the fish has grown and as its skeleton formed. I have also had others comment on this particular phenomenon in the past, but again, research into their cause has not got too far. However, most of the scientists who have considered these swellings suggest that they may be a reaction to a parasite, bacteria or virus infection from an early age. This seems like a pretty good bet to me, although as far as I know, no-one has sectioned the swellings and looked at them under a microscope. I would tend to lean towards a

A classic "old man snapper" showing the head hump and fleshy lips. The question is, why do these features occur in some areas and not others?

parasite of some kind, and would also predict that fish from some areas are commonly infected (like, from southern Sydney for instance) while those from other areas are not (no proof of this yet).

Now, if you have been following this logic, you won't be surprised that I would now suggest the swellings on the heads of snapper might well be the result of a similar (or identical) infection within the skeleton of the fish. Certainly, this would explain why snapper from some areas are free of humps (no infecting organism), and why humped snapper seem to only occur over reefs (the infecting organism is associated with reefs in some way – perhaps an intermediate host is a reef-dweller too). The theory may even explain the fleshy lips and nose of old men snapper, being an external reaction to the infection later in life. I realise that this may be a somewhat controversial theory, but in the absence of any others, it's certainly worth a small research grant to a budding student. I for one will certainly be following any ensuing research with great interest.

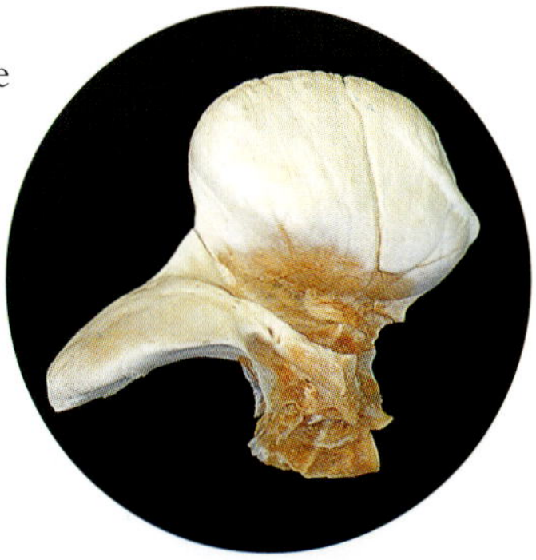

The skull of the snapper pictured earlier. Note the massive swelling of the cranium. Does it have a purpose, or is it due to external factors?

The Big Sting

I have heard of anglers having some pretty nasty experiences in relation to stings from marine creatures such as fish, bluebottles and the like, and I've had my fair share of memorable encounters too. However, this story takes the cake in that department. Fishing in the surf at the north end of Stockton Beach, near Newcastle, early in the morning, our fisherman felt a searing pain on his ankle. He noticed a pale blue jelly-like substance on his leg, which was too soft to remove easily, but by then, the pain had become excruciating. His mates helped him to his 4WD, only 50 metres away, where he tried to rest, but within a short time, his breathing rate had increased to about 100 per minute, and his temperature was "sky high". A few minutes later, his breathing became slower and slower, to the point where he thought he was fading away. His mates rushed him to the nearest medical centre where his blood pressure was found to be very high (although he had never suffered from high blood pressure before). A local anaesthetic was injected near the welt on the ankle and, after a while, his blood pressure and temperature returned to normal. He was allowed to leave, but the intense sting lingered for 18 hours before it finally went

All jellyfish can sting but, fortunately, most are relatively harmless.

away. Even over the next five days, he had a high temperature and bouts of sweating; altogether a most unpleasant and frightening experience.

It left me wondering what it could have been. From the symptoms, my initial suspicion was a box jellyfish (a sometimes deadly marine stinger), but to make sure, I decided to consult a medical colleague who is one of the leading authorities on marine stings, bites and poisons in Australia, if not the world. Dr Carl Edmonds specialises in diving medicine, and has written an excellent book entitled *Dangerous Marine Creatures*. (I highly recommend this book to all anglers, since it contains a wealth of information on all forms of stings, injuries and poisons derived from marine animals, including symptoms, first aid and medical treatment – often very helpful for the less well-informed medicos.) Dr Edmonds didn't take long to come up with a confident diagnosis. The offending stinger was almost certainly a box jellyfish, based on the symptoms of rapid, followed by slow breathing, temperature reversals and on the description of the soft jelly-like substance on the sting site – a characteristic of this stinger. It turns out that there are several species in this group of jellyfishes, known as Cubomedusae, some of which are subtropical in their distribution. There is even a small, relatively harmless Cubomedusoid residing in Sydney Harbour, which was news to me. Therefore, while the Stockton Beach stinger was not necessarily the common tropical variety that keeps swimmers out of the water in our north throughout summer, because of the severity of the symptoms, it was concluded that this particular one may indeed have been a stray from the tropics which had been blown down by cyclonic weather. Apparently, stings from box jellyfish that far south on the east coast are very rare (but more common at this latitude in WA).

In areas where it occurs, such as northern Queensland, the box jellyfish, or seawasp, needs to be taken very seriously.

Proper Identification

Scientists, and medical practitioners in particular, are very keen to make positive identifications of animals causing such traumatic, potentially fatal effects on their victims. Most people are aware of the importance of correct ID in the case of spider and snake bites, since they realise that this information may help treatment, and even avert death. They therefore usually make an effort to catch (or kill) the offending animal and take it with the patient to the nearest hospital or medical centre.

A familiar sight to Australian swimmers, the bluebottle, Physalia physalis, *packs a painful sting, but is rarely, if ever, fatal.*

The same is true for marine stingers and other nasties. If possible, put the animal (fish, jellyfish, coral, worm, shellfish, etc) in a jar or plastic bag, or in the case of jellyfish, a part of the tentacle will be sufficient for pinning down the type of stinger. If the organism is identified, treatment will probably be much more effective. For box jellyfish, an anti-venom has been developed and is available in northern Queensland; not much help in the south since death, if it occurs, is normally within an hour at the most of the sting.

FIRST AID FOR STINGS

As far as immediate first aid of stings is concerned, some treatments are quite effective and bring some immediate relief. For example, the excruciating stings from the spines of venomous fish such as stonefish, fortescues, bullrouts and catfish can be relieved by applying heat in the form of hot water. (The nearest thermos flask or car radiator is a sometimes forgotten source.) For bluebottle and other jellyfish stings, including mild ones from box jellyfish, douse the area with vinegar and lift off the tentacles carefully. The vinegar stops discharge of the tiny stinging cells, whereas rubbing (an old remedy) causes many more of these cells to fire into the skin. Applying anaesthetic sprays or creams to the site may also help for both fish and jellyfish stings. Most of the old treatments for stings such as dousing with alcohol or even petrol, and cutting or applying tourniquets, are now discouraged.

At times, the marine world may seem an unfriendly place, filled with all kinds of stinging, biting and poisonous creatures, but these seemingly unfriendly characteristics have evolved to provide either food or protection to their owners. It's just that we humans, out of our natural environment, sometimes get in the way.

Death, Where is Thy Sting?

Swimming, snorkeling and diving can sometimes seem pretty dangerous past-times. Apart from the perceived dangers of being bitten, or even eaten, by predators ranging from sharks to barracudas, care must also be taken when touching many species of fish, molluscs, crustaceans, corals, jellyfish, and even worms. (I well remember taking a walk out on a reef flat at low tide, and spotting a small, marine worm – a bristle worm, or polychaete. I bent down to gently touch it, and for my troubles, received a painful volley of tiny glass-like spines in my hand. So much for a simple stroll).

But the two main culprits which cause marine stings are jellyfish and fish. The latter cover a wide species range, including members of the scorpion fish family (such as red rock cod, lion fish, fortescues, stonefish, etc), spadefoots or rabbit fish, surgeon fish, gurnards, catfish, and even our local flatheads. The other large group of stingers, the jellyfish, is an amazingly diverse group which all tend to have one thing in common: tentacles which can impart stings, ranging from mildly uncomfortable to excruciating, and sometimes, even lethal.

Venomous and Poisonous Marine Animals: A Medical and Biological Handbook, edited by John Williamson and others, and published by UNSW Press, is a massive undertaking. It covers the medical literature on all forms of marine poisoning.

Rabbitfish (happy moments, spinefoot)
Siganus fuscescens
40 cm
1.6 kg

Temperate to subtropical waters of eastern and western Australia. The rabbitfish is ironically known as "happy moments" due to the intensely painful sting which it can inflict with its dorsal fin spines.

Stingrays can and do cause excruciating injuries, but only two known deaths in Australia have been attributed to stingrays.

Because there are so many different kinds of jellyfish, much of the book is taken up with medical descriptions of the symptoms and treatment of stings from this broad group. To the layperson, the detail would probably be a bit daunting, but to the medical profession, this work is a major contribution. Many GPs just don't have the time, let alone the resources, to gather such information, so I would hope that this book finds its way onto the shelves of a great many local doctors, especially those with practices near the coast.

One of the fascinating topics covered in the book is determining if actual fatalities have been caused by marine animals in Australia, and in other countries, and if so, how many. The authors exhaustively searched the medical literature and other sources, and came up with some relatively unexpected results. For example, they found that the number of confirmed deaths caused by many marine animals in Australia was far fewer than popularly believed, especially for those feared critters, the stonefish, blue-ringed octopus, cone shells, sea snakes and bluebottles. On the other hand, deaths from jellyfish stings proved to be more common on a worldwide basis than had previously been realised.

The main killer among the jellyfish is, not surprisingly, the box jellyfish (actually a number of related species), hundreds of fatalities being ascribed to them, mainly in the Philippines, Indonesia and, of course, Australia. In Australia, there are 61 deaths attributed to box jellyfish, and while this is considerably less than the 180-plus deaths recorded as caused by sharks, the authors make the point that far greater numbers of people are badly injured by box jellyfish than by sharks. (In other words, jellyfish encounters are more common than shark attack, but fortunately, result in proportionally fewer deaths.)

The box jellyfish, Chiromex fleckeri. *More deaths are attributed to this small invertebrate than any other marine stinging animal.*

Looking at deaths in Australia definitely caused by other marine animals, we learn that there have been only two caused by the blue-ringed octopus, two by stingrays (in both cases, the barb puncturing the heart), one from a cone shell, one from ciguatera poisoning, and 14 from eating puffer fish. And just to round off this rather morbid list, there have been 27 recorded deaths from crocodiles since the 1870s.

In other parts of the world, animals other than jellyfish that are known to have caused fatalities by envenomation include: sea snakes (responsible for at least 150 deaths per year), cone shells, blue-ringed octopus (but only one other fatality outside Australia), sea anemones, stingrays, stonefish (only three deaths on record), catfish, weever fish and scorpion fish (butterfly cod).

Having revealed that the box jellyfish is the most dangerous critter in our waters, what is the latest on initial treatment of stings from its long, trailing tentacles? There is a whole section in the book on this topic, and space doesn't allow me to go into much detail. However, the basic drill is to call immediately for an ambulance, pour vinegar over the affected area and apply compression bandages to the major stings (but only after at least 30 seconds of vinegar dousing). Don't waste time trying to remove tentacles that have been well doused in vinegar, and perform CPR if necessary.

One final comment on this excellent book. It contains a lengthy section of full colour photographs, many of which show the injuries caused by stings from many different kinds of jellyfish and other animals. This section is not for the weak stomached out there, but the photo that made me really wince is one showing a Japanese diver with the head of a large longtom (needle fish) deeply embedded in his eye socket. The poor chap lost his eye, but amazingly, survived this shocking encounter.

Even though death by lethal weapon is rare in the marine world, encounters of the stinging kind can be extremely unpleasant. As with all things in nature, take care, and when in doubt, don't touch.

Poison Poisson

To eat or not to eat? That is the burning question that confronts many northern Australian anglers whenever they land a good-sized mackerel or reef fish. And the reason for the gnashing of teeth (or in this case, not gnashing)? Ciguatera. This odd word (from the Spanish cigua, meaning a marine snail which causes poisoning) refers to a weird phenomenon whereby a wide range of tropical fish species may develop quite poisonous flesh. For many years the cause was a mystery, since not all fish of a species are poisonous, while some individuals of normally edible species may occasionally cause the characteristic poisoning.

The symptoms of the poisoning are also wide ranging and difficult to diagnose. They include nausea, vomiting, cramps, and strange sensations such as heat reversal (hot objects feel cold, and vice versa) and loss of feeling in the feet. Symptoms can return for many years after initial exposure, especially if fish containing a trace of the toxin are eaten again.

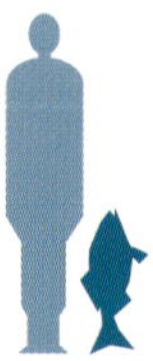

Coral trout
Plectropomus leopardus
75 cm
16 kg

A tropical reef species through most of the northern half of Australia, as far south as northern New South Wales on the east coast. Large coral trout have a relatively high association with ciguatera poisoning, especially on the Great Barrier Reef.

After considerable research worldwide, it was discovered that the toxic substance, which accumulates in the muscle of affected fish, comes from single-celled blue-green algae living on tropical reefs. (Related blue-green algae are the cause of those toxic blooms, which sometimes affect inland rivers.) The algae are grazed on by small animals, which in turn are eaten by larger fish, and so on up the food chain. Larger, older fish keep accumulating the poison, and so are most dangerous. Presumably because of their diet, some species of fish are very prone to being affected, and therefore are a definite culinary no-no in the north. These include red bass and Chinaman fish. Large mackerel and coral trout are, unfortunately, also affected, as I can vouch for.

Once in Townsville, I woke in the wee small hours with incredible pains in my legs, intensely itchy palms and soles of the feet, and nausea. It wasn't until the next evening, when the symptoms returned after a cold beer, that I realised I had been struck by the dreaded ciguatera (alcohol brings on the symptoms again) and thinking

Another prime candidate for ciguatera poisoning, the Chinaman fish, Symphorus nematophorus. *Queensland specimens are treated with caution, but fish from Western Australia have been given the all clear.*

back, I had eaten a large meal of coral trout 12 hours earlier. I was told later that mine had been only a minor encounter, and if that's true, my sympathies go to anyone having a worse experience. Which brings me to the following points about ciguatera poisoning.

Often diagnosis is stumbled upon by one of the doctors the patient is referred to – the point being there are many people concerned about the ignorance and the lack of knowledge in the medical profession concerning this poisoning and its effective treatments (if any). A large proportion of the population lives on the coastal fringe north of Coffs Harbour, where the incidence of the poisoning is probably greatest. There are also, in all probability, many Australians with medical problems as a result of ciguatera poisoning, some with minor effects, some with major effects, and many probably also undiagnosed.

Again, I turned to one of the leading authorities on this subject in Australia, Dr Carl Edmonds. Dr Edmonds has devoted a whole section of his book on marine medicine to ciguatera poisoning. It details symptoms, treatment and avoidance of the problem, but I wondered myself whether medicos in general would be well versed in these erudite contents.

Dr Edmonds agreed that it is very likely that many cases of ciguatera poisoning go unnoticed, or at least, undiagnosed, in Australia each year, but pointed out that this was to be expected. Firstly, many of the ciguatera symptoms are so variable, and react so differently to the known forms of treatment. Only recently it was discovered that there are at least three, and very likely more, distinct toxins in the family of ciguatera poisons and that these all have different effects. Dr Edmonds agrees that obviously not all medicos would be able to make an instant, positive diagnosis, but he also makes the important point that it is impossible to be expert in all branches of medicine, especially toxicology. Marine biologists, after all, can rarely speak with authority outside their specialised areas, and many of them would know very little about ciguatera. I take the point, but I am sure that if all GPs north of Coffs Harbour were aware of the problem, a lot more cases would be diagnosed and treated earlier.

A Long Way from Home

A knock on the door early one morning revealed a wet colleague standing on the doorstep in full fishing gear holding a bucket. He had been fishing off the rocks south of Sydney and had picked up a strange little silver fish driven from the water by some lurking predator. There in the bottom of the bucket was a pretty little fish about the size and colour of a 20 cent piece, but its identity was a bit of a puzzle. I had seen similar sorts of fish in South East Asian fish markets, but headed for my fish identification books to make sure. The little lost fish turned out to be, as I suspected, a pony fish, probably the pugheaded pony fish, *Secutor ruconis*, a member of a large family of tropical, shallow-water fishes which abound through northern Australia and South East Asia. But what was it doing so far from home?

This is a classic case of what is known in the trade as an expatriate fish; quite literally a fish that is "away from its fatherland". The limits or boundaries of the known distribution of a species of fish define its range. In virtually every fish book one cares to consult, the distribution and range of each species is always described, often by means of a map. But there are always exceptions to the rule.

In my early days working at NSW Fisheries, I had many enjoyable days in Sydney Harbour aboard the small research trawler *Port Jackson*. The project involved a study of the diversity of fish found in one of Australia's most urbanised estuaries and involved systematic trawling at specific sites throughout the harbour each month. Apart from revealing much about the common fish species of the region, an unexpected offshoot of the study was the finding of well over 500 species of fish – many more than would be expected at this latitude. The reason for this was the occurrence, sometimes quite commonly, of many species of tropical fish not

A tropical pony fish, Secutor ruconis, *from Sydney. A fish out of water in more ways than one.*

Tropical reef species, such as the well-named pineapple fish, Cleidopus gloriamaris, *are occasionally carried to the temperate waters of Sydney Harbour.*

normally associated with temperate habitats. As well, because of the fine mesh of the net used, many of these tropical visitors were present as very small juveniles. Closer examination of the data, and later plankton studies, have shown that every year, eggs and larvae of many tropical species are carried down the coast from northern Queensland with the warm East Australian Current. Most of these tiny outcasts inevitably die, but some find a nook, a cranny, a fin-hold, and eke out an existence as true strangers in a strange land. However, because the cooler conditions are not suitable for their maturation and eventual spawning, these expatriates will die without leaving any progeny behind.

Often, large adult fish dwell at the edges of their species' range. Bermagui is a case in point, where unusually large specimens of many tropical pelagic fishes have been caught over the years. These include the all-tackle Australian record yellowfin tuna, dolphin fish to 40 kg, big wahoo, 300 kg blue and black marlin, and a real surprise, the all-tackle world record longtail tuna. This area on the far South Coast of New South Wales tends to be about the southernmost limit of the tongues and eddies of warm water which extend down the east coast in early autumn. The theory has it that large specimens of species, which prefer warm water, being able to tolerate cooler water because of their surface area to volume ratio, skirt the boundaries between water masses to avoid competition from their smaller relatives.

But back to the pony fish. This little fish appeared off southern Sydney in mid September, suggesting an unusually early intrusion of warm water from the tropics well down the coast that year. At about the same time, another fishing mate had mentioned catching some scaly mackerel at Solitary Island, New South Wales, as well as, wait for it, several juvenile dogtooth tuna. Scaly mackerel (also called large scaled tuna or double-lined mackerel) are a well-known species caught on the Barrier Reef and used for marlin bait, and as I discovered, have sometimes been recorded from Solitary Island in the past. But the appearance of dogtooth tuna that far south, according to the Australian Museum, is most unusual.

It should also be said that not all expatriate fish found off eastern Australia are the result of one-way traffic from north to south. Occasionally a southern ocean species turns up well to the north of its usual home range. A case in point was an eel caught off Botany Bay. For a positive identification of the fish to be made it was taken to the Australian Museum where the staff identified it as a southern conger eel, normally found off Victoria and Tasmania, this time an unusual northern record for another expatriate fish, a long way from home.

The scaly mackerel, Grammatorcynus bicarinatus, *is usually associated with northern Australia, but sometimes may wander a long way south.*

A Pink Bolt from the Blue

No matter how much we might think we know about the marine life in our waters, there's always something that pops up now and again which shakes us up and gives us a big surprise. This amazing fishing story begins with Nick Botic, member and club champion of the Fairfield RSL Small Boats Fishing Club, fishing in a club competition off the Sir John Young Banks, about 100 km south of Sydney. He had been trolling for yellowfin with a Pakula cockroach lure and was working his way back to port. He had intended to make a final flathead drift when he noticed a school of large slimy mackerel under the boat. Quickly, he hooked one, and thought, why not bridle the bait to the lure hook for a last troll? You will no doubt have guessed that the combo lure/bait was promptly taken, and Nick settled down for quite a tussle. In fact, after a while, he passed the rod around to his fellow anglers all the time thinking that they had hooked a yellowfin tuna. After about 75 minutes, and with Nick back with the rod, the double appeared and colour showed. Instead of the silver and yellow flash of a yellowfin, an eerie pinkish orange colour reflected up from the deep. As the fish was brought closer to the surface, everyone's eyes started to bulge. This was a good sized fish all right (80 kg in fact), but the strangest fish our intrepid crew had ever laid eyes on. With some difficulty, the monster from the deep was boated, but still had enough kick left to land on top of Nick and leave him covered with bruises. But any discomfort was forgotten in the elation, and in asking the big question: what on earth was this weird and wonderful fish?

The first I heard of the capture was a telephone call later that day from a friend of Nick's, John Spence. "Have you ever heard of a yellowfin tuna with the head of a dolphin fish?" he asked. My first thought was of a deformed tuna (not unheard of) but when Nick called the next day and gave me a few more details, especially the pink-orange colour of the fish, it was immediately clear to me that this was a rather large specimen of the rare and mysterious louvar (also called luvar and luvalu).

Anglers show off the catch of their lives – the weird and wonderful louvar, Luvarus imperialis.

So what, you may well ask, is a louvar? It's a large pink-orange fish with a small mouth, powerful tail and a blunt head, indeed reminiscent of a dolphin fish.

Above: The rare and enigmatic louvar. A large, mid-water fish about which virtually nothing is known.

Below: The incredible goblin shark, Mitsukurina owstoni, *another large yet rarely seen marine animal.*

But where does it live, what is its biology and how common or rare is it? Delving into just about every possible source of information I could muster led mostly to blanks. However, it is known that the louvar (scientific name *Luvarus imperialis*) is the only member of its family, Luvaridae, and occurs in temperate to tropical waters worldwide. Most known specimens have been found washed ashore, and it is generally accepted that this is an open ocean, mid-water species that spends most of its time well away from land. Its bright colouration is similar to other strange fishes of this twilight world such as the oarfish and the opah, also suggesting that it dwells in relatively deep water (orange and red appear black in the depths). Studies of juvenile and larval specimens of louvar (also rare) suggest that the most likely close relatives of the louvar are the surgeon fishes (much smaller reef-dwelling fish equipped with razor sharp "blades" on their tail wrist). This is particularly interesting since one of the other groups of large oceanic fishes, the huge sunfishes, are closely related to the toadfish. Two cases of small, inshore fishes with giant, sea-going cousins.

One question this capture raises is how common are louvar off Australia? A call to Mark McGruther, an ever-helpful colleague at the Australian Museum, revealed that only four specimens exist in their collection: three of these were caught or washed up in the 1990s, but one, caught off Southwest Rocks in 1939, was much larger. This particular fish was also referred to by Arthur Parrott in *Sea Angler's Fishes of Australia* (1959), who gives its weight as 225 pounds (102 kg) and length at 6 ft 4 inches (193 cm). Parrott also mentions that the first record of a louvar in Australia was washed ashore at Bermagui in August 1901, but does not give its size.

One of many species of surgeon fishes, common in the tropics. Scientists consider the louvar to be most closely related to the surgeon fishes.

Coincidentally, I recalled seeing a mention of another rod and reel capture of a louvar in the American magazine *Sport Fishing*. Sure enough, the magazine carried a short piece about a 50 kg louvar caught by a game fish angler off Baja, California, the same day that he landed a striped marlin. As the article noted: "To catch a louvar on hook and line is truly a rare experience."

The existence of the louvar – a substantially sized but rarely encountered fish – serves to remind us that there is still much to learn about the marine world. Like the louvar, other large denizens of the ocean about which very, very little is known are sometimes caught or washed ashore. The weird alien-like goblin shark is a case in point, as is the aptly named deep-sea filter-feeding mega-mouth shark. Both very large sharks, but only known from a few specimens. We may think we know a lot about the nature of marine life, but every now and then, the oceans reveal that we still have a lot to learn.

Colossal Calamari

One of my earliest memories of a movie is surprisingly vivid. The scene is a huge, writhing squid attacking the good submarine Nautilus, with Captain Nemo hacking off squirming tentacles as one of the ship's crew is slowly lowered towards a huge parrot-like beak. (Perhaps that scene is what made me want to be a marine biologist, but somehow I doubt it.) The movie, of course, is *20,000 Leagues Under the Sea*, based on the famous Jules Verne novel.

My mind naturally went back to that scene when I picked up the classic novel recently for a re-read. Written in 1870, the book tells of the adventures of an incredible futuristic undersea vessel exploring the abysses of the world. Of course, the famous squid attack scene is the highlight.

The humble squid, Sepioteuthis lessoniana, *is an awesome predator in its own right. But its giant namesake would be a sight to behold.*

What prompted me to re-read the novel was an article in the magazine *Seafood New Zealand* on giant squid. This article, by government biologist Steve O'Shea, pointed out that there are, in fact, several species of squid that grow to impressive sizes (2 to 3 metres in total length). One he colourfully describes as "one seriously evil, actively carnivorous monster ... seizing and restraining its prey with wicked, sharp, talon-like hooks set in each arm, and then shredding the live prey with its parrot-like beak".

However, only one species can truly lay claim to the title Giant Squid. A huge version of its smaller cousins, *Architeuthis dux* is the largest invertebrate (animal without a backbone) in existence. How big? Well, the largest specimen of this incredible creature so far encountered was one measuring 18 metres, which was washed up in New Zealand in the late 1880s. While the length includes the very long pair of feeding tentacles, it was still estimated that this colossal calamari weighed at least a ton, or 1000 kg. And some squid scientists think that the species may even grow to a length of 23 metres.

As is usual for deep-dwelling creatures (they are thought to live at depths of 200 to 800 metres), very little is really known about its biology. The little information that has been gleaned has come mainly from stranded specimens, or the few individuals that have been trawled at great depths. However another unusual source of information comes from the study of sperm whales. It appears that these largest members of the toothed whale family rely very much on giant squid as a staple of their diet. During the days of whaling, routine examination of sperm whale stomachs commonly found remains of giant squid, either chunks of muscle and tentacles or, more often, the large beaks which are not easily digested. Many sperm whales also showed saucer-sized scars from the suckers of giant squid, usually around the head.

The sperm whale/giant squid connection has led to an ambitious project. Run by the Smithsonian Institute in Washington DC, the idea is to attach cameras to sperm whales, and hopefully capture live giant squid on film. This sounds like an expensive, rather fanciful exercise, but I would imagine that the resulting footage would be in high demand (although I have the rather ironic thought that the images may look just like small squid, with no scale or object with which to compare them).

One of the most incredible magazine articles I've ever read appeared in the March 1941 issue of that most famous of magazines, *National Geographic*. Entitled "Fighting Giants of the Humboldt" by David Duncan, it recounts an expedition off the coast of Peru, led by one of the originators of game fishing, Michael Lerner. The purpose of the adventure was somewhat scientific, but the sampling tools were game fishing rods and reels. During a month-long trip, the expedition caught many marlin and some very large broadbill swordfish. This in itself was ground-breaking stuff, but the pinnacle of the story was the discovery while fishing at night that they could actually catch giant squid on game fishing gear. The particular species they encountered is not identified, but the accounts of their capture make for fascinating reading. With heads covered by pillowcases as shields against the jets of ink, there was no trouble enticing a squid to take the offered baits. In Duncan's words: "No sooner did the bait fall, amid its shower of phosphorescence, than the rod whipped and snapped as line faded from the reel. The moment we dropped a new rig overside the rod arched again as the line was yanked one way, then another. The reel howled. First it was full, then empty, then full, then empty again." After finally gaffing one, Duncan continues: "Only then did we realise how grotesque and truly fearsome were these creatures we had been fighting. That first squid caught on rod and reel was a terrible thing. One end supported 10 armlike tentacles, which were frightening to behold. Nestled at the base of these tentacles was a glistening beak, against the slashing attack of which few living things could survive."

To me, the amazing aspect of this encounter was that this squid was only 3 metres and 50 kg. With descriptions like this and the one above, it is hard to even contemplate what a 1000 kg or larger one would be like. Jaws author Peter Benchley must have had the same thoughts as he penned his pot-boiler about a 100-foot long, rampaging squid, simply titled The Beast. I missed the ensuing four-hour TV mini-series, which may be just as well – my vivid childhood memories of the original classic can now remain undisturbed.

SEQUEL

Before I wrote this piece, I had only heard reports of rare specimens of giant squid being trawled up from the depths. However, no sooner had I finished writing it than I had a call from Bermagui charter captain Lloyd Williams. Lloyd had been out on his boat *Esprit* when he saw an almighty commotion on the surface of the water. Moving closer, he was surprised to see a fair sized mako shark thrashing about, apparently involved in a tussle with what looked like a large squid. Sure enough, after the shark departed, the remains of its meal were hauled aboard, and the remains would have made a heck of a lot of calamari. Only the tubular body of the squid was left, but this was at least the size of a 43 kg yellowfin tuna.

Scientists from Museum Victoria examine a specimen of the giant squid, Architeuthis dux, *caught accidentally in a trawl net off Portland, Victoria. Much larger specimens are known.*

Amazingly, about the same time I read a newspaper report of a giant squid about 3 metres long caught by a trawler off Portland, Victoria. And to top it off, colleague John Ashley also related his own sighting of a colossal cephalopod. John spotted his on the surface off the east coast of Papua New Guinea a couple of years ago. When the boat drew near, the squid, also about 3 metres long, darted away in a cloud of ink. Perhaps these monsters of the deep aren't so rare after all.

And Thereby Hangs a Tail

Just when I'm starting to wonder what's new in the fishing world, along comes a fascinating story from my colleague Lester Rohrlach in Lae, Papua New Guinea. Lester loves his fishing, but even more, loves thinking about and studying the fish he catches.

"Fishing in a vast school of thousands of jewfish over a 300 metre front of the ocean bed off one of my favourite rivers near Lae," says Lester, "and in a sampling of only six dozen there were three fat, healthy and as equally strong fish that had lost the tail and peduncle right up to their second dorsal fin in the long, long ago (perhaps only a year or two in fishy terms). Another had lost 10 per cent of its body along the top, including all but a crooked leading spine of the former first dorsal fin. One had lost only half of the tail fan. Peanuts, that. They were survivors, having learnt to compensate with their body movement and trailing dorsal fin for propulsion both to compete with their fellows to secure food ... and to elude their adversaries."

The fish in question were tropical relatives of our more familiar jewfish, or mulloway, the assumption being that the tails were bitten off by predators (sharks, mackerel, or whatever) and that the fish lived to fight another day. The interesting point was that the proportion of tail-less fish was actually quite high, at 4.2 per cent of his catch, and, although the thought crossed his mind that the lack of a tail might

A catch of Papuan jewfish, Johnius vogleri, *showing a couple (right) which had lost their tails, presumably in an earlier near-death encounter with a toothy predator.*

be genetic, he tended to lean towards the escaped-from-the-jaws-of-death hypothesis.

Bearing in mind that the fish had different parts and amounts missing, I would agree that these individuals were very likely victims of near fatal encounters with toothy carnivores. I guess the odd thing was that the predators, having nipped off the tails, did not come back to finish the job, as is their usual modus operandi.

Lester Rohrlach with a tail-less jewfish from Lae, Papua New Guinea. Living proof of survival of the fittest.

Thinking back, I once collected several bream that completely lacked a tail from a commercial catch, but they were almost certainly deformed from hatching as larvae. In those cases, the lateral line continued right around the end of the body where the tail would have been. Also, as Lester suggested for his jewfish, the dorsal and anal fins had taken on the role of the tail, or at least partly so. In that case, the most likely explanation was a deformity during development of the embryo when the tail simply never formed at all. In fact, birth defects in fish are almost certainly quite common, but in the fish-eat-fish world of survival of the fittest, the great majority would die at an early stage so we rarely actually see them in the catch.

Survival from injuries and/or deformities has been an intermittent theme in my travels and writings. I've written about 15 kg yellowfin tuna with markedly bent spines, even larger yellowfin completely missing the long anal sickle fin and pugnosed southern bluefin tuna, all of which were likely to be genetic deformities. Injured fish, which obviously survived, included a fair sized jewfish with only one eye, apparently from an old hooking injury, a whole swag of species with chunks missing, and perhaps the most amazing of all, a blue marlin with a completely healed gill arch that had been torn away at the base during a previous encounter. In the last case, the original injury would have been the equivalent of having one's carotid artery severed, but again, the fish survived.

A conference in the US that I was fortunate to attend was entirely devoted to catch-and-release in marine recreational fisheries. Many papers were presented on

Even though the loss of its tail would have been very serious, the wound has completely healed and the fish appears to have compensated well.

survival of fish after hooking, the point often being made that some species are more robust than others. For example, slides were shown of halibut (giant flounder) with incredible injuries to their jaws caused by previous hooking which had completely healed, while small salmon were particularly prone to internal injuries from hooks (to the extent that circle hooks are now mandatory in some areas).

Tagging programs constantly confirm that at least some fish survive traumatic injuries: the recaptures of these now perfectly healthy fish, on checking the tag cards, show they had been apparently mortally wounded at release. Comments on the cards such as "bleeding badly from the gills", "hooked deeply", "hooked in eye", "not expected to survive" and so on would seem to be death sentences. And yet, the recapture of these fish after many months or years with no obvious signs of their earlier injuries is living proof of survival.

Blast from the Past

I've always been a sucker for second-hand bookshops. In fact, the more second-hand, the better. That musty aroma wafting up the nostrils when one enters the premises is unmistakable, and the thrill of the chase (or should that be, thrill of the browse?) is hard to beat. And it's not hard to guess which shelves I target – straight to the "Natural History", or even better, "Marine Life", sections, of course. Not that those categories are usually very well stocked, but every now and then, a rare find can really make the day. So it was that during a visit to a particularly good Melbourne bookshop my eye was drawn to the broad spine of a real treasure. Titled *Creatures of the Sea – Being the Life Stories of Some Sea Birds, Beasts and Fishes* by Frank T Bullen, published in 1904, a quick flick through its handsome pages meant there was no escape for my wallet.

Unfortunately, unlike modern books, there is no biography on the author, but it is obvious in reading the narrative that Frank T Bullen was a pretty amazing sailor and fisherman. He recounts his fishing adventures aboard many vessels all over the world, from England to the Caribbean and the Bay of Bengal. He even makes a brief reference to when he was a "lamp trimmer" on a small steamer running between Sydney and Grafton, catching what he thought were scaleless "baracoutta" but which were almost certainly Spanish mackerel.

What a great read this treasure of a book is. Written in that rather grand, formal style of the day, but still quite chatty, it is one man's account of his love of all living marine animals, and his desire to pass on that information to the public.

While many of the observations and descriptions are surprisingly insightful and accurate, Bullen occasionally wanders into fanciful territory. For example, his account of swordfish attacking whales through sheer aggression is somewhat over the top but nevertheless fascinating. It is almost certain that swordfish and marlin impale whales entirely accidentally while chasing fish, but Bullen thought that they not only attacked whales out of blood-lust, but even killed baby bowhead whales for food.

Collecting old books on marine life can be fun – and even educational if you're willing to learn from the past.

"To bear away, impaled by one swift and terrible blow, the newly born calf from beneath its mother's protecting arm is a supreme delight, and one too that yields solid results in the shape of food, for the tender body of a young whale is easily divided by repeated blows of the sword."

One of the interesting writing methods he uses is to tell the story from the point of view of the animal in question. For example, there is a long narrative "written" by a sperm whale recounting its life cycle and adventures, including its near fatal encounters with whaling boats. This approach is also used to good effect in his account of the life of the "albacore", which I suspect is really the Atlantic bluefin tuna. (This method of outlining the life cycle from the perspective of the fish reminded me of the same sort of treatment in a famous book written some years ago by David Scholes called *Trutta the Trout.*)

In one of my favourite passages Bullen takes a bit of a swipe at academic stay-at-home scientists. He writes: "But a great difficulty confronts the museum naturalist in the study of the deep-sea fauna. He is so greatly dependent upon stray specimens brought home for his observations, he cannot, in the nature of things, bring personal experience to bear, and he must theorise largely. Which of course tends to error, since it is, to a great extent, working in the dark."

While he often waxes lyrical about the bounteous ocean, Bullen still seemed to have had a remarkably good grasp of natural fluctuations in fish numbers. In a fascinating passage on the sardine fishery off Breton, he writes that porpoises were thought to have caused reductions in their numbers, to the extent that ... "a paternal government has, at the request of the fishermen, granted the use of torpedo boats to hunt and destroy the porpoises in the hope that thus will the drain upon the supply of sardines be stopped. But it seems doubtful indeed whether such measures are of the slightest use. The natural checks placed upon their [sardines] increase in the sea by the Creator will never do more than they should do, when unmolested by man. It is only when man thrusts his hands into Nature's work that any great unbalancing of natural resources takes place."

"Albacore" (probably bluefin tuna) feeding on flying fish. Some early naturalists were extraordinarily accurate observers.

(Now why don't they write like that any more?)

Sometimes we get pretty smug in these days of instant information overload, so it is good to be brought down a peg or two now and then by realising that not all knowledge has emerged in the last decade or two. All too often, we dismiss earlier work, and go on to reinvent the wheel. So next time you're strolling past that old bookshop, poke your head in the door and follow your nose, and curiosity. You never know where it might lead.

The copious bounty of the sea, in this case, blue mackerel. Bullen was aware, though, that there must be limits to how much can be caught.

Voyages of Discovery

It's been said before, but we often take a lot of things for granted. Like going to sea in modern vessels, like using sophisticated scientific equipment, and like having ready access to a wealth of information on fishes accumulated over many years of study. Of course, things weren't always like that. Two hundred years ago, the natural world was a place of great mystery. Lands and seas over the horizon were only dimly known and the early seafaring explorers set out on their voyages with great trepidation. On board some of these ships, however, were naturalists who must have been incredibly excited at the prospect of discovering new and bizarre animals, plants, and, of course, fish.

Many of these early voyages were multipurpose in nature, in particular those of Captain James Cook. Most of us know that Cook's first voyage that brought him to *Terra Australis* was primarily mounted to observe the transit of Venus across the sun from the best vantage point near Tahiti. Most people would also have heard of Joseph Banks, who was a passenger on that first voyage. But more than a passenger, Banks was a VIP gentleman scientist whose fare contributed substantially to the costs of the voyage. Banks was an eminent naturalist, and was in charge of the Endeavour's scientific program. Among other things, this involved recording, and often collecting, specimens of most of the plants and animals found, including, of course, fish.

Illustrations of a snapper, Pagrus auratus *(left), and an Australian salmon,* Arripis trutta *(right), from Cook's first voyage. In the absence of cameras, such illustrations were essential scientific recordings, but also stand as aesthetically beautiful works of art in their own right.*

Cook made three monumental voyages across the Pacific, and on each of these, naturalists and artists were an integral part of the ship's company. I must admit that I was very much unaware of the work of the artists on Cook's voyages until I came across a wonderful book entitled (deep breath) *Forty drawings of fishes made by the artists who accompanied Captain James Cook on his three voyages to the Pacific, 1768-71, 1772-75, 1776-80, some being used by authors in the description of new species.* Now that's a mouthful, but with a title like that, well, I just had to have it. Published by the British Museum of Natural History in 1968 to mark the bicentenary of Cook's first voyage, it contains a portfolio of beautiful reproductions of just some of the fishes these intrepid scientists and illustrators faithfully recorded throughout the Pacific.

There were four artists on these voyages whose work is illustrated, and each has his own style. Some fish are sketched in pencil, some are simple unfinished watercolours, and others are finely detailed works, but what really struck me was the accuracy of the depictions. All fishes are immediately recognisable, even though the Latin names of the day have been radically revised since.

There were 270 paintings and drawings of fishes from Cook's voyages, and 37 of those in this publication had never been published before; they had only been seen by those privileged enough to have access to the collection in London. The fishes

selected for the portfolio are a mixed bag including such familiar species as the shovel nosed shark, banjo ray, oxeye herring, Australian salmon, snapper, silver drummer, stargazer, jackass morwong and gemfish.

Apart from the scientific motivation for collecting fish, the obvious nutritional value of fresh protein dictated the need for catching fish. No doubt there were some pretty good anglers on board those early voyages, but then again, just imagine being able to dangle a hook into completely virgin waters. The mind boggles.

But it wasn't all milk and honey, and as a reminder of the hardships these pioneers endured, I was touched to learn that two of the illustrators on Cook's first voyage died before reaching home again, probably from scurvy. Nevertheless, their legacy remains in the form of the first depictions of some of our most familiar fishes. And of course, if it weren't for the foresight of the Trustees of the British Museum of Natural

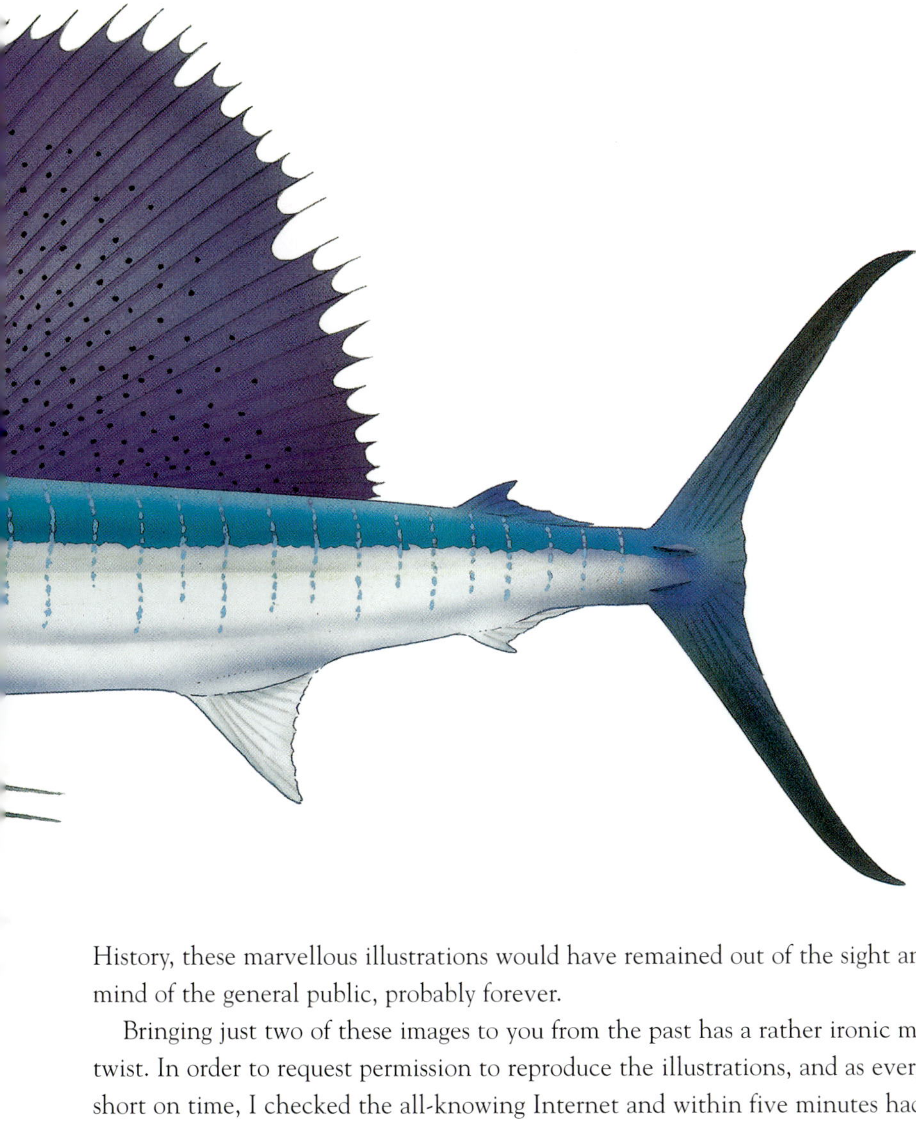

Even today, when all manner of photographic recording devices are available, only a drawing can convey the true beauty of the sailfish.

History, these marvellous illustrations would have remained out of the sight and mind of the general public, probably forever.

Bringing just two of these images to you from the past has a rather ironic modern twist. In order to request permission to reproduce the illustrations, and as ever, being short on time, I checked the all-knowing Internet and within five minutes had found the website of the British Museum of Natural History. A brief e-mail to the address given was answered within 24 hours, graciously allowing reproduction of the illustrations, for which I am most grateful. I suspect those intrepid artists and naturalists might have raised a wry smile if they knew just how small the world was to become in 200 short years, but at the same time knowing how their work, produced under extraordinary conditions, would certainly stand the test of time.

Biodiversity and Fish

The biodiversity of life on this small, blue planet of ours is perhaps nowhere better illustrated than in its seas, oceans and rivers. One only needs to don a face mask and snorkel, and dive in to gain a sense of wonder at the incredible variety of fishes which populate just about every aquatic habitat on Earth. There are more species of fish than all other vertebrates combined, and their variety is, quite simply, mind-boggling.

Tropical reefs are home to an amazing varitey of fish species

The Diversity of Life ... and Fish

I don't have enough time to read these days, well, read for entertainment, that is. Funnily enough, when I do pick up a paperback at the airport on the way to or from meetings (or fishing trips), I tend to go for one of those popular science titles that seem to be proliferating of late. And I've been reading just such a book lately – *The Diversity of Life* by Edward O Wilson – and a fascinating book it is. This is an account of the enormous range of the forms of life on earth, in the wake of a worldwide push for the recognition of the importance of biodiversity. The term "biodiversity" simply refers to this variety of life, but it now has taken on the meaning of the interwoven nature of life, and the importance of that diversity to life's continuation on this planet.

Turretfish
Tetrosomus concatenatus
30 cm

Found along the coasts of New South Wales, Queensland, Northern Territory and Western Australia. Variety of habitats including reef and seagrass. Related to pufferfish.

Thumbing through this rather scholarly work, I was struck by the chapter on the number of species of animals and plants currently existing on earth (only a fraction of the species that have ever existed). Before going into that, however, a short lesson in biology may be called for. The term "species" means a population of animals or plants that can freely interbreed with each other but not with individuals of other groups or populations. In other words, the members of a species (never specie, by the way) form a reproductively isolated group which exchanges genes amongst themselves but not with outside groups. In the past, those who classified living things into species had to rely on external features to make the distinctions – things

There are more species of fish than all other vertebrates combined. Just one of the 20,000 plus species, the blue ribbon eel, Rhinomuraena quaesita.

like numbers of fin rays, measurements, body parts and so on. Nowadays, however, the powerful tool of genetic DNA fingerprinting has revolutionised the art of classifying animals and plants, and can be used to clearly differentiate between and among species.

According to Wilson, the approximate numbers of known species on earth are as follows:

bacteria – 4800; fungi – 69,000; algae – 26,900; higher plants – 248,400; protozoa (amoeba, etc) – 30,800; sponges – 5000; corals, jellyfish, etc – 9000; flatworms – 12,200; round worms – 12,000; earthworms, bristle worms, etc – 12,000; molluscs – 50,000; starfish, urchins, etc – 6100; insects – 751,000; crustaceans, spiders, etc – 123,400; fishes – 18,800; amphibians – 4200; reptiles – 6300; birds – 9000; mammals – 4000.

Apart from the extraordinary number of insect and crustacean species, the other aspect of this list that impressed me was that fish, with nearly 19,000 species, make up almost half of all known species of vertebrates, that is, animals with backbones. (A recent encyclopedia of fishes on my shelves states that there are 24,000 species of fish, the discrepancy only illustrating how little we really do know about life under the water.)

The variety, or biodiversity, in the world of fishes is truly remarkable. Consider, to begin with, the sharks and rays. Now sharks and rays are widely thought to be quite primitive – sort of living fossils – and while it is true that they existed before the bony fishes and their body forms have not changed a great deal over millions of years, they have, in fact, evolved into just about every niche imaginable in marine waters. Take, for example, the smallest shark – the dwarf lantern shark, measuring only about 20 cm at full maturity – while at the other end of the spectrum the extraordinary whale shark reaches a size of at least 12 metres and weighs more than 12,000 kg.

Now consider the so-called bony fishes, or teleosts, so named in contrast to the sharks, whose skeletons consist of cartilage rather than bone. With 24,000 plus species, it is not possible to go into detail, but looking at just a selection of different forms serves to illustrate their incredible diversity. Lungfishes; sturgeons; saratogas; eels (many varieties ranging from one species only 4 cm long to the morays, and the giant conger eel); herrings, sardines and pilchards (the largest herring is the tarpon, which grows to over 150 kg); carps (a diverse family including the humble goldfish and the mighty mahseer of India which grows to perhaps 3 metres); carachins, which include many aquarium fish such as the tetras; catfishes, from the huge European wels as large as 5 metres and 300 kg, to walking fish, to the infamous candiru which sometimes swims up the urethra of humans; salmon and trout, which need no introduction; angler fishes, including some truly bizarre deep-water varieties; cods, which feed millions; the amazing flying fishes; grinning lizard fishes; seahorses and pipefish, including the master of camouflage, the leafy sea dragon; scorpion fishes, including the deadly stonefish and the spectacular lionfish; groupers, including our own formidable Queensland groper; the colourful wrasses; the many and varied gobies and blennies; flounders and soles and puffer fish, which include the sunfish – perhaps the largest of the bony fishes. And then there are the perches, all 9000 species of them, including such diverse species as butterfly fishes, snapper, jacks, tuna and marlin. If this brief outline captures your imagination, you could do worse than to grab a copy of *Encyclopedia of Fishes* (UNSW Press) edited by John Paxton and William Eschmeyer.

The truly bizarre porcupine fish, Diodon nichthemenus. *There are more species of fish than all other vertebrates combined.*

It is a sobering thought that the figures of species quoted above refer to the number of species actually discovered and described by science. The true number of species on the planet is thought to be many times this; quite possibly as high as 30 million.

What's in a Name?

The names of fishes can cause considerable consternation amongst anglers, commercial fishermen and scientists alike. As any travelling fisherman would know, the common name for the same species of fish can change drastically between geographic locations. Take the small tunas and mackerels, for example. A spotted mackerel can be called a snook, doggie mackerel or spotty depending on the regional dialect. A bonito (or bonita) in northern Queensland can be a mackerel tuna (or is that a horse mackerel or little tuna?) or a striped or skipjack tuna elsewhere. The freshwater yellowbelly changes its name to a callop when it swims down the Murray into South Australia. Cross international boundaries and the name game becomes even more mystifying. A tailor here is a bluefish in the US is an elf in South Africa is a tassergal in West Africa. The common names can even appear as antonyms, as is the case for the black marlin. A simple enough name, except that the Japanese name for the same fish is Shirokajiki, which literally means white marlin.

The list of confusing common names of fishes is of course practically endless, but these examples serve to show that somewhere along the line, it is important for some consensus to exist as to the proper name of a given species, so that everyone can be in some agreement regarding which fish is under discussion, or under the microscope. In the United States, attempts have been made to standardise the common names of fish, that is, to elect an official common name for each species. This has not been attempted in Australia, although there are mandatory marketing names for many species to protect consumers. We therefore must rely on the scientific names of fish to avoid confusion here.

The bestowing of names on flora and fauna would seem to be an inherently strong human urge. Anthropologists have theorised that the act of knowing an animal's

Bonefish
Albula vulpes
100 cm
8.5 kg

Inhabits reef flats, lagoons and sometimes estuaries in the tropics, throughout northern Australia and islands of the Indian and western Pacific oceans. A legendary sportfish, its scientific name makes sense to anglers. It means "white fox".

name transfers some control over the animal to the name-giver or name-holder. And so it seems to be with fishermen, and perhaps even more importantly, fisheries biologists and ichthyologists.

Early classifiers of the living world struggled with naming systems for many years. Towards the end of the 17th century, long, descriptive names in Latin were being used. Thus, the humble carnation plant was called *Dianthus floribus solitaris, squamus calycinis subovatis brevissimis, corollis crenatis*, meaning pink, with single flowers, the scales of the calyx somewhat egg-shaped and very short, with scalloped petals. Obviously something had to be done. Fortunately, a Swiss botanist by the name of Linnaeus entered the scene in the mid 18th century and devised a simple, workable system for classifying and naming all organisms. He strived for a natural order and suggested that all animals and plants could be given a unique, double-barrelled, or binomial, Latin name. The first name would signify the genus, and the second the species. The genera (plural of genus) could then be arranged into families, families into orders and so on, ending with the animal and plant kingdoms. Under this new system, the carnation would simply be called *Dianthus caryophyllus*, and no other species could be given this name. (By the way, the Latinised name is always printed in italics, with the first name, or genus capitalised, but not the second, species name.)

The mulloway, or jewfish, has been given a real tongue-twister of a scientific name, Argyrosomus hololepidotus *– a necessary evil in avoiding confusion over identity of species.*

The ingenious system of Linnaeus was quickly accepted by the scientific community, and is the only accepted international nomenclature system in use today, called, appropriately, the Linnanean system. Even in scientific papers written in non-Latin root languages, such as Japanese and Arabic, the internationally recognised binomial Latinised name of a given animal or plant must be used.

Now that we know the origins of the Latin names of fishes and other animals, it is interesting, not to mention entertaining, to look at a few of the better known species and translate their picturesque names, often bestowed many years ago.

The legendary bonefish and the broadbill swordfish were named by Linnaeus himself. He aptly called the bonefish *Albula vulpes* or "white fox", and the broadbill, Xiphias gladius, the two words both meaning "sword", one in Greek and one in Latin. The striped marlin, *Tetrapturus audax* roughly means "bold, with a four-winged tail" (referring to the two pairs of keels on the tail wrist) while the sailfish *Istiophorus platypterus* means "sail-bearer with flat wing". The tailor, as mentioned, has many common names throughout the world, but is always known scientifically as *Pomatomus saltator*, from "serrated operculum" (gill cover), and "one who dances or leaps". Another species with a rather romantic name is the remora, or sucker fish. Called *Echeneis naucrates* by Linnaeus, its name refers to the legendary power of the sucker on its head, and means "a pilot, which holds a ship".

Some Latin names may sound evocative, but their English translations are not so romantic. Many of the flatheads for example, belong to the genus *Platycephalus*, which means, not surprisingly, "flat-headed", while the common yellowtail is called *Trachurus* meaning "rough tailed". The mulloway or jewfish has been stuck with a real tongue-twister, *Argyrosomus hololepidotus*, but this simply means "silver bodied, with smooth scales". Some names don't seem to capture the true essence of the fish. For example, the threadfin salmon, a great fighting fish in many people's books, is called *Polynemus plebeius*, meaning "many threaded, common, or vulgar". Other names may even seem a little insulting, for example some members of the "sole" family are dubbed *Cynoglossus* which means "a dog's tongue" (although I can see why).

Many Latin names of fish refer to their old common names in other counties, and their true origin has been lost. The yellowtail kingfish is *Seriola lalandi*. "Seriola" is an old Italian name for this group of fish, while "lalandi" refers to Delalande, an early naturalist after whom the species was named.

I am attempting here to demystify the nomenclature of fishes and help make the system more friendly. All too often, the joys of science can be clouded through the use of technical terms or jargon. Taking the time to find out the meanings of names of fish can be fun, and downright educational.

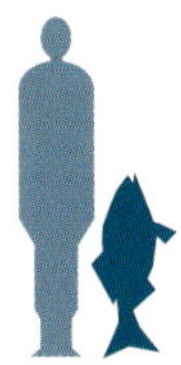

Dusky flathead
Platycephalus fuscus
110 cm
12 kg

Primarily an estuarine dweller, from southern New South Wales to central Queensland. The dusky flathead is an ambush predator, whose name rather unimaginatively translates as "flat head, dusky".

Cosmopolitan Fish

One of the fascinating things I find about travelling to other countries is being able to visit the local fish markets, or even better, an aquarium displaying local fish. It's always exciting to see new and strange species (at least, strange to what one is used to), and considering that there are over 24,000 species of fish in the world, the chances of coming across unfamiliar ones in faraway places is obviously high. This is especially true for freshwater fish on other continents, and also for shallow-water coastal fish. Given the huge distances separating such areas in different continents, the evolution of very different forms of fish in widely separated areas around the world is to be expected. However, what I've always found surprising is the fact that some species of fish occur over very broad ranges, across entire oceans, and even more spectacularly, throughout the whole world. A case of seeing familiar faces in the crowd.

Very few species of plants or animals other than fish have been able to spread around the whole planet (without help, that is). Fish, or at least a few highly successful species, seem to be an important exception to that rule. Before outlining some of those species, let me say that we are not talking about the entire globe here. No one species occurs from the Poles to the Equator, since temperature ranges are critical for all species of fish. Rather, worldwide species usually occur within certain temperature bands in all of the three major oceans (Indian, Pacific and Atlantic). These distributions are sometimes termed circumtropical, or circumtemperate, but there are some species, the southern bluefin tuna is one, which remain mostly in the cooler latitudes of the three oceans. Therefore, the term "cosmopolitan" has been coined to describe species that occur right around the globe. This is not perhaps the best word, since it doesn't really convey what is meant, but it is a useful one and it is here to stay. The barriers that obviously prevent movement of many fishes between oceans are the continents themselves, but even these obstacles have not prevented the spread of some species.

A wahoo, Acanthocybium solandri, *photographed in* Vietnam. *A few species such as this are distributed in all of the major oceans of the world.*

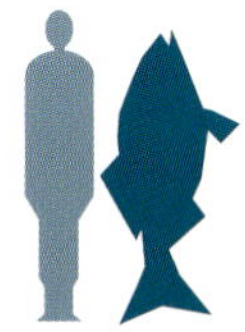

Opah (moonfish)
Lampris guttatus
180 cm
275 kg

The moonfish is a true cosmopolitan species, occurring in all three major oceans. It is primarily a mid-water species which is rarely encountered, except by longline fishermen. Its flesh is delicious, and is known as "salmon of the gods" in French Polynesia.

By far the vast majority of cosmopolitan species can be classified as pelagic, or surface-dwelling fishes. This is to be expected to some extent, since the surface layers of the ocean are constantly moving in predictable currents – the so-called "ocean wind". We know from drift buoys (and, for that matter, messages in bottles) that these currents can and do move floating objects over long distances. In fact, colonisation of islands by plants, and some animals, occurs by such means (the seeds of certain plants are highly adapted to this mode of dispersal). Obviously, oceanic currents will also transport fish larvae and eggs, but tagging programs have also shown that adult fish may also move over very large distances, cross oceans, and even move between oceans.

Because there are so many cosmopolitan species, I will only list the main ones that we might be familiar with. The major group missing from this list are certain deep-water fishes such as lightfish, gulpers and others, which are not encountered by anglers.

Of the billfishes, only blue marlin, sailfish and the broadbill swordfish could be called truly cosmopolitan since only these three species occur in all three major oceans. In the case of tunas, true worldwide species are the yellowfin, albacore, bigeye, southern bluefin, skipjack and frigate mackerel. And of all the small pelagic bait-fish, only the slimy mackerel seem to have attained worldwide status.

Other worldwide pelagic species include the dolphin fish (actually two worldwide species), wahoo and cobia. Cosmopolitan fishes are also found among mid-water

The cobia, Rachycentron canadus. *A worldwide species, and the only member of its family,* Rachycentridae.

species such as the sunfish, opah, oarfish, ribbonfish, snake mackerel, oilfish, and Ray's bream (or pomfret).

Many of the surface-dwelling sharks are cosmopolitan, maybe because they have been around for a long time. Most widespread are probably the mako and the blue sharks. Other world citizens are the thresher, tiger, bull whaler, oceanic whitetip whaler, and the great hammerhead. In size contrast, the small cookie-cutter shark is spread worldwide, so successful must be its odd, parasitic lifestyle. The large plankton-eating, basking, mega-mouth and whale sharks, are also distributed throughout the major oceans, as is the largest of the manta rays. One particularly interesting species of shark in this respect is the Great White shark. It is certainly found all over the world, but it is entirely coastal and has a very patchy distribution. The big question is, how, if at all, do individual white sharks move between these areas?

Cosmopolitan coastal species are much less common than pelagics. The bonefish is a good example, occurring near islands and atolls in the three major oceans. And if you use a tailor rig in South Africa or the eastern US, you will be likely to catch fish, which, to all intents and purposes, are tailor (although their local names will be different). Another seemingly universal coastal fish is the good old bully or sea mullet (the striped mullet of the US). It is found all over the world but, curiously, it is the only mullet that is.

The big question, of course, is how did this come about, and why are a few species so widespread, while other closely related fish have quite restricted distributions? Although we don't really know the answers to these questions yet, the science of population genetics, using DNA fingerprinting as a tool, is slowly unpicking the lock. It turns out, for example, that tailor populations from the eastern US have quite different DNA fingerprints from those taken off eastern Australia. This indicates slow divergence over time, but as yet not enough for the development of separate species. So next time you are in another country, pop into the local fish market. After all, it's a small world, and you just might spot an old friend.

Warm-blooded Fish

Like most kids of my day, or any day for that matter, I was completely enthralled by dinosaurs. I knew all their names, where they lived, how big they were, and made regular visits to museums to gaze at the skeletons and reconstructions of how they might have lived. In those days, dinosaurs were considered by scientists to be basically big reptiles – slow, dim-witted and lumbering – and like all reptiles, they must have been cold blooded. Over the past couple of decades, however, there has been a revolution in the way we think about these fascinating animals. By piecing together snippets of information from many sources, it is now fairly well accepted that many of the dinosaurs were almost certainly warm blooded, and therefore were able to remain active at night and in winter. As well, they were probably very nimble and agile, as portrayed so well in the movie Jurassic Park, and in the television series *Walking with Dinosaurs*. It seems that dinosaurs were in fact closer to birds in their internal make-up than the primitive reptiles, and it may well be that birds are really surviving dinosaurs.

But what does all that have to do with fish? Well, again, the widely accepted view has been that fish, being even more primitive than reptiles (and amphibians for that matter) are all cold blooded, their body temperatures being completely controlled by the temperature of the surrounding water. But in the words of the old Gershwin song, "it ain't necessarily so".

The black marlin, Makaira indica, *maintains a body temperature slightly above that of the surrounding water. It also has a brain heating organ behind the eye.*

Just because fish have been around for a long time doesn't mean that they haven't evolved, and indeed, we only need to take a flick through any fish book to see how diversified fish have become in their adaptations to the watery world. And while it is true that the great majority of the 24,000 or so species of fish on the planet are cold blooded, the exceptions to the rule are pretty amazing.

TUNAS

Of all fish, the tunas as a group have evolved the most advanced form of temperature control. All of the tunas are able to maintain their core body temperature above that of the water in which they swim, some more than others. Skipjack tuna, for example, may have a core temperature up to 15°C higher than the surrounding water. The heart and blood vessels of tunas are structured so that heat is conserved or shed as needed. This can depend on their activity, the water temperature and the size of the fish, smaller fish needing to conserve heat more than larger ones since their surface area to weight ratio is higher. The heat the tuna traps in its blood and body tissues is, of course, generated by digestion and by muscular activity. In other fishes, this heat is continuously lost to the surrounding water, whereas the tunas and some other species are able to control this loss by means of complex networks of heat exchange blood vessels. Now, this form of body heat control is by no means as efficient or constant as that of mammals or birds, both of which maintain body temperatures within very narrow ranges. The body temperature of tuna, and the other fishes mentioned later, still fluctuates fairly widely, but is not at the mercy of the surrounding water temperature to the extent of that of the great majority of fish.

BILLFISH

The marlins also maintain elevated body temperatures, but not to the extent of the tunas. Blue marlin tagged with sonic tags and tracked off Hawaii showed a slightly higher muscle temperature than the surrounding water, but we also know that core temperatures of marlin can be up to 10°C higher than at the body surface.

The broadbill swordfish is also classified as a billfish, but is not really very closely related to the marlins. Unlike marlin, broadbill dive to great depths where water temperatures can be as low as a few degrees above freezing. To cope with such extremes, the broadbill has developed an amazing organ behind the eye that produces heat and warms the brain and eye at the same time. This organ, known as the brain heater, is actually a modified eye muscle which can no longer contract, but instead produces considerable excess heat. This heat is channelled around the brain and the retina in order to keep the fish alert and able to react quickly in the cold depths, obviously giving the awesome predator a big advantage over its cold-blooded prey such as herring and squid.

Contrary to the belief that all sharks are primitive, some species, such as the mako, Isurus oxyrinchus, *have evolved the ability to control body heat to some extent.*

The amazing thresher shark, Alopias vulpinus, *is also somewhat warm blooded.*

Although the broadbill has the largest brain-heating organ, marlin and some of the tunas also possess similar, smaller organs. These are thought to have evolved separately in each of the three groups of fish, indicating that this is obviously a very handy piece of equipment to have tucked away behind your eye if you happen to be an oceanic predator.

Sharks

Perhaps the real surprise package in the world of warm-blooded fish comes in the shape of some sharks. The popular conception of sharks is that of cold-blooded killers, but the shark with the worst reputation, the Great White, is actually nothing of the sort. Whites, and their cousins, makos and porbeagles, can maintain body temperatures between 4°C and 10°C above surrounding water temperature. It is also thought that thresher sharks have this capacity too. These two families of sharks are generally regarded as the most highly evolved of all sharks, and even though their external body shapes have changed little over hundreds of millions of years, their internal biology certainly has.

Odds and Cods

A story that made headlines around the world was the rather gruesome tale of the human head found in the stomach of a cod caught off the Barrier Reef. This was not an urban myth but a true story, and although I won't dwell on the morbid details of the much-publicised incident, it did generate considerable interest in cods in the media and among the general public. Many questions were asked about cods, including how many species there were, how large they grow, and whether they pose a threat to humans.

The question of how many cods there are is impossible to answer, simply because "cod" is a sort of generic name given to so many different types of fish. The Northern Hemisphere cod is, of course, the prolific species that was dramatically overfished last century, and about which a best-selling book, simply titled *Cod*, was written.

I have noted the confusion caused by using common names, which can change from coastal town to coastal town, and the name "cod" is probably the best example of this in Australian fishing. A quick look at the index of just a few of my books on Australian fish shows just how hard it can be to work out what a particular cod is just by referring to its common name. Ern Grant's *Guide to Fishes* is a good place to start, especially because he fished extensively in Queensland, and was very good at his reef fish identification. Grant lists no less than 57 cods in his index, which is mind boggling to say the least. Many of these are secondary common names, but nevertheless it does indicate the range of species with this moniker.

Generally speaking, the term "cod", at least on northern Queensland reefs, most often refers to members of the genus Epinephelus. The Epinephelus suite of species is a large group of closely related fish belonging to the family Serranidae, and are mostly

One of the numerous species of Australian "cod", the flowery cod, Epinephalus fuscoguttatus. *Overseas, the group is known as groupers.*

A beautiful old painting of one of the "cods", the Queensland Groper, Promicrops lanc-eolatus – *enough to give a kid the willies.*

called groupers in the US. Familiar examples of these would be the wire-netting cod, flowery cod, potato cod and tomato cod (great names, aren't they). Also in this genus are the black cod and estuary cod, both of which are protected in New South Wales. Many other less familiar cods are mentioned by Grant, including the speckle-finned cod, Morgan's cod (wonder who he was), maori cod (not to be confused with the maori wrasse), comet cod, blackspot cod, blue-tailed cod, black-saddled cod, long-finned cod, blackfin cod (also known as spotty cod – naturally), white-dotted cod and others – but I think you get the picture. As noted, all of these cods have the first Latin name Epinephelus, but naturally (and inevitably) there are exceptions to this rule. The best example would have to be the famous Queensland groper, which is also an Epinephelus species, but not called a cod. More on the name "groper" in a little while, though.

You will note that most of these cods have fairly simple, descriptive names, which leads me to relate a rather amusing story. Each year, a fun fishing tournament event is run on Hamilton Island, Queensland. It is primarily a game fishing competition, but a few years ago it began to include a reef fishing section for smaller boats. The trophies for the tournament are magnificent painted plates depicting the various species eligible for weigh-in, and in one particular year there was a special category for the heaviest orange-spotted cod. I was present at the weigh station to identify all fish, and armed with copies of the best books, thought that things would be fine.

One of the larger cod species, the potato cod, Epinephalus tukula.

How wrong one can be. Fish after fish was duly presented to me for the category "orange-spotted cod", but the problem was, every one of them was different.

Checking the experts, I could identify white-spotted, yellow-spotted, plain old spotted, birdwire, and even black-arsed cod (truly), but nothing matched what was supposed to be the "official" orange-spotted cod. At the presentation dinner, the beautiful platter was consequently awarded to the local identity (who shall remain nameless) who had originally suggested the category. Funnily enough, orange-spotted cod was missing from the eligible species list the next year.

Well, that's probably enough confusion for now. However, I did mention the name "groper" that is also used rather loosely around our shores. This name almost certainly derives from the name "grouper", which is a common name for a lot of reef fishes in the US. In fact, the famous Nassau grouper is very closely related to our own equally famous coral trout. And having mentioned trout – no, I think that's enough said on the subject.

Writing all this has reminded me of one of my vivid childhood memories. It was a painting in my dad's well-thumbed copy of *Roughly's Fish and Fisheries of Australia* of a huge Queensland groper lurking on the bottom with mouth agape. I was convinced that a fish like that could easily swallow a man, let alone a child, and for some reason that image used to give me nightmares. The recent gruesome incident in Queensland brought those images back, but now that I know cod are harmless, well, they don't bother me at all – right? Well, maybe just a little bit.

Dragons, Ghosts and Horses

Most fish look like – well – fish! You know, head, tail, scales and the usual set of fins in roughly the same, familiar positions. But as we have learned on these pages, adaptation and evolution in fishes has led to an amazing array of diverse lifestyles and body shapes. And right up there among the most bizarre of all is the endearing seahorse.

Most people would think of seahorses as those little, classic coil-tailed creatures encased in their spiky armour, inhabiting a sort of underwater fairyland. Charles Kingley's Water Babies rode on them, and seahorse caricatures feature in just about every underwater cartoon ever made. It may therefore come as a surprise to learn that many different kinds of seahorses populate the underwater world. All belong to a single family containing over 200 species worldwide, of which, at least 110 species call Australia home. This odd family, named Sygnathidae, contains not only the familiar seahorses, but an intriguing menagerie including pipefish, pipehorses, and sea dragons. (The equally remarkable ghostpipe fishes belong to the closely related family, Solenostomidae.)

The extraordinary, and well-named weedy sea dragon, Phyllopteryx taeniolatus. *Now under protection in New South Wales, this species is only found in Australian waters.*

One of the more unusual species within the seahorse family, perhaps the most spectacular of them all, was recently brought to my attention by a photo sent to me by a colleague. It was a delightful creature he found freshly washed up on a central New South Wales beach after a major storm, and wondered about its identity and possible rarity. With such a clear photo, it wasn't hard to identify this truly remarkable fish as a common or weedy sea dragon (*Phyllopteryx taeniolatus*), one of only two sea dragons in the seahorse family (the other being the leafy sea dragon). Both of the sea dragons are native only to Australia, and both are the ultimate masters of disguise. Their skin is festooned with seaweed-like tendrils and fronds, perfectly mimicking the colour and texture of the plants in which they live. The common sea dragon

Adult (above) and juveniles (right) of probably the most amazingly camouflaged fish in the ocean, the leafy sea dragon, Phycodurus eques.

belies its name to some extent, since it is now protected in New South Wales waters. This precaution is not so much because of proven population declines, but because it is a prime target of aquarists. The sea dragon now has such a following in marine circles that there is even a "Dragonwatch Database" maintained by the Marine and Coastal Community Network.

The ghostpipe fish, Solenostomus paradoxus, *a wonderful example of the diversity of fish shape and form.*

Even though they may appear weird to us, the seahorses and their cousins have obviously adapted extremely well to their mode of life. Camouflage is largely the secret to this success, their method being to resemble as closely as possible underwater plants, including many types of seaweed and seagrasses, each species taking on the colouration and textures of their preferred habitat.

Not long ago, while dragging a small seine net over a seagrass bed in Botany Bay, I was absolutely amazed to discover the numbers of pipefish living in such a small area. In a two-minute tow using a 10 metre net, we caught literally hundreds of pipefish, perfectly camouflaged to look exactly like strands of eelgrass. Under normal circumstances, they would completely escape notice, and only became obvious when the contents of the net were emptied into a temporary aquarium on the beach.

While their body shapes vary a lot, a common feature of all of these fishes is their long tube-like snout, which is used effectively to pluck passing zooplankton from the water. For this precise method of feeding, they need stereoscopic vision, and it is always a joy to watch their eyes swivelling forward as they home in on their tiny targets.

Having said that most members of the seahorse family are camouflage experts, it should be no surprise that some inevitable exceptions to the rule exist. For example, two brightly coloured members of the group, the cleaner pipefish and the banded pipefish, are coral reef species, which have evolved the garish colouration of cleaner wrasses and cleaner shrimps respectively. Cleaner fish and shrimps are brightly decorated presumably to signal to larger fish that they are available to provide the service of picking off parasites from the skin, gills and mouth, and not being targeted as prey in the process. Both of the mimic pipefishes have also evolved this life style,

Not all the seahorse family are masters of disguise. The cleaner pipefish, Doryrhamphus janssi, *(the two slender fishes), signal their role to larger fishes.*

The familar shape of the short-head seahorse, Hippocampus breviceps.

closely copying the colouration of the more common cleaners which the host fish are used to. The host fishes are none the wiser and the pipefishes are able to pinch their free feed.

Seahorses and their relatives are not harvested commercially in this country, although they are very popular in the aquarium trade. However, there is a market for seahorses in Asia, which has rung some warning bells. Dried seahorses are apparently in high demand for so-called traditional medicines, to the extent that concerns are held for some species. A venture in Tasmania, which aims to breed and farm seahorses, has received a lot of publicity. The proponents of the project are targeting this market, ultimately with the aim of satisfying the demand for seahorses with their farmed product, thereby protecting wild stocks. This sounds very noble, but it remains to be seen if farmed supplies can meet the ever-growing demand. Let's all hope it's not just a pipe(fish) dream.

Pickled Fish

I've always liked museums. Not just on rainy days, and not only when a new exhibit comes to town, but anytime that the mood strikes, a stroll around the cabinets and displays will always reveal some new and interesting fact. But museums aren't only built for educational purposes. Behind locked doors and usually in cluttered basements, scientists work continuously on describing and understanding life on this planet in all of its forms. Museums are the repositories of collections of just about anything you can imagine, and like the proverbial iceberg, what you see on display is only a small fraction of what is stored below.

Several years ago, I had the chance to visit the mother of all museums – the Smithsonian Institute's American Museum of Natural History in Washington DC. The Smithsonian is actually a series of many museums covering everything from history to art to natural history and I took the opportunity to visit the latter after attending a conference in Monterey, California.

Washington is still a long way from downtown Monterey, but the magnet which attracted me there was a display being developed on the life cycle of the black marlin. This display became possible with the donation of the cast of the still-standing, world record black marlin, caught off Peru in 1953 by the donor, Alfred Glassell Jr. The museum staff decided to make the most of it, and started working on a permanent exhibit showing the growth of marlin from larvae the size of a pinhead to the true leviathans of the deep. I had come into the picture by earlier arranging for a cast of a very small juvenile black marlin to be sent over for the exhibit. The juvenile was

Museums are not only for display, they are also research centres and repositories of biodiversity. This preserved stargazer, Kathetostoma laeve, *is the first of its kind to be described and catalogued.*

only 50 cm long, and was caught off Jervis Bay, south of Sydney, in 1991 – the smallest black marlin ever caught on rod and reel.

After helping out with distribution maps and other marlin facts, and having been highly impressed by the thoroughness of the research for the display, I was then asked if I would like to see the fish collection of the Smithsonian. That was like asking my son if he really wants a day off school to go surfing?

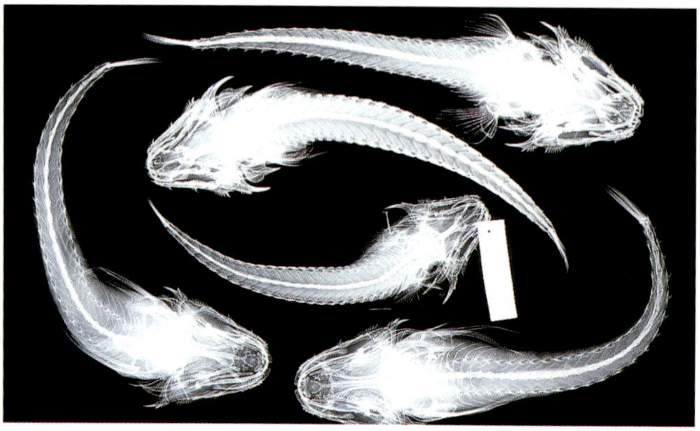

Specimens of tiger flathead, Neo-platycephalus richardsoni, *tagged, pickled and X-rayed for posterity.*

In great anticipation, I was led down passageways and through musty vaults under the museum, until finally I was ushered into a vast room with row after row of packed shelves disappearing into the dimly lit distance. Each shelf was crammed with neatly arranged bottles of every size and shape containing preserved fishes in all of their incredible diversity. There were specimens from all over the world, dating back to the 18th century when the natural world was only beginning to be sorted and classified by European scholars. Here were the so-called "type" specimens of thousands of species of fish – that is, the first specimens of their type to be collected and used to describe in detail, complete with fin ray and scale counts, the species for the first time. I was told that the collection was the largest in the world, containing over 500,000 lots (a lot might be one fish, or a whole collection from an expedition, some of which are still awaiting expert sorting). Even the containers were incredible. The older specimens were all in beautiful sealed glass jars, many of which had been especially manufactured for a particular fish. This really was ichthyology heaven, and I was even able to inspect some tiny sailfish and blue marlin of which I had only previously seen photographs.

Living fossil

But the real highlight of the day was yet to come. I was taken over to a long stainless steel tank and when the catches were all undone and the lid removed, there it was perfectly preserved in its bath of ethanol – a real honest to goodness coelacanth. Now a coelacanth is to a fish biologist what a living dinosaur would be to a reptile expert. It is a genuine living fossil, virtually identical to ancestors found preserved in rocks nearly 400 million years old, and all fish biologists know the story of its discovery. The first living coelacanth was discovered in 1938 by a young biologist, Marjorie

Courtenay-Latimer, when a fisherman showed her an unusual fish he had caught off the coast of South Africa. She immediately knew that this was no ordinary fish, and arranged for it to be taken to the South African museum. There it was confirmed that this was a Crossopterygian, or lobe-finned fish, which were the ancestors of all land animals, and which were supposed to have been extinct for at least 70 million years.

The shelves of museums hold the treasures of the past, and often the keys to future discoveries.

Fittingly, the fish was named after its discoverer – *Latimeria chalumnae*. Since then, quite a few coelacanths have been caught off the Comores Islands, near Madagascar, living specimens have even been filmed underwater, and recently, some have even been found off Indonesia. Perhaps not surprisingly, they move their lobed fins just like a four legged animal, and it is not too difficult to imagine them strolling up onto some ancient beach to begin the long, slow evolutionary journey toward, well, you and me.

Next time you're going past a museum, take the time to pop in and stroll around the many and varied exhibits. You will certainly learn more about some of the wonders of life, and you won't be disappointed.

Big Fish Down Under

Fishing stories abound with tales of the big one that got away. And most of the time, the truthfulness of the teller of the tale is brought into question, if only via a raised eyebrow. But sometimes, such stories are true. There is something truly awesome about a really big fish! The fishes of the open ocean, the marlins, tunas and some sharks, may all grow to enormous sizes, but our knowledge of them is remarkably poor. Gradually, though, the jigsaw puzzle is taking shape.

The black marlin is an awesome fish by any standards

Of Baby Bills

One of the more significant billfish captures of the century was made off the New South Wales coast in 1991. No, it wasn't a world record blue marlin, or a line-class broadbill, but a specimen much more important to the eye of the biologist – it was the smallest billfish of the family Istiophoridae (the marlins, spearfishes and sailfish) ever taken from New South Wales waters. The chain of events leading to this, and the final identification of the specimen, unfolded a bit like a detective story.

It started for me with a telephone call from Tim Simpson of the *Compleat Angler*. He had been aware of my search for small billfish for years, and when one of his staff told him he had seen one caught, Tim knew it could be significant. The little billfish, all of 50 cm long, had taken an 18 gram Juro Baitfish lure cast from the rocks inside Jervis Bay, New South Wales. The two anglers involved in the catch were alert enough to realise they had something pretty unusual, so they carefully transported it to a local fish-moulder, Fred Montebello. He immediately made a light cast (to minimise damage) then folded all the fins down, and popped it in his freezer.

When Tim telephoned me about the fish, he said that it could be a little sailfish. I was excited at this prospect, but knew from past experience not to jump the gun. There had been a spate of very small (1 to 2 kg) broadbill swordfish caught off New South Wales several years ago, and I was not discounting this possibility on this occasion. Anyway, I rushed down to Fred's at Callala Bay and watched in anticipation as he lifted the package from the freezer. Sure enough, and to my absolute delight, as he lifted the protecting cloth back, he revealed a beautiful specimen of a true billfish.

I immediately knew it wasn't a broadbill, since it had two pairs of keels near the tail for a start (broadbill have one pair) and it had small pelvic fins under the chin (broadbill have none). But which of the marlins – spearfish or sailfish – was it?

A gorgeous miniature black marlin, Makaira indica, *showing the very high dorsal fin characteristic of all baby billfishes.*

The initial suspicion by everyone who had seen it was that it was a baby sailfish. It was long and thin, and had a high sail-like dorsal fin. Yes, it did look like a sailfish, however, the height of the dorsal didn't seem right to me. I recollected that small sailfish, about the size of this one, had been caught in the Atlantic Ocean in the Gulf of Mexico, and these had all had very high, arched dorsal fins. This one was much flatter. (All of the billfishes have sail-like dorsal fins when young, but with the exception of the sailfish, the body outgrows this fin as they get older.) The flatter dorsal still didn't completely negate a sailfish, since very small Pacific Ocean specimens had not been properly described before. The other candidates in the line-up were striped marlin, Pacific shortbill spearfish or black marlin. (The blue marlin could be safely discounted because young ones of this size from both the Pacific and the Atlantic oceans have virtually no bill, and a very obvious "chicken wire" patterned lateral line.)

Now, the scientific literature on juvenile billfish is anything but extensive, so there was no recipe to look up in a "What-Fish-is-That" type book. Many of the features distinguishing adults from one another (such as relative height of the dorsal fin, etc) are useless for juveniles, so I had to look for more subtle clues.

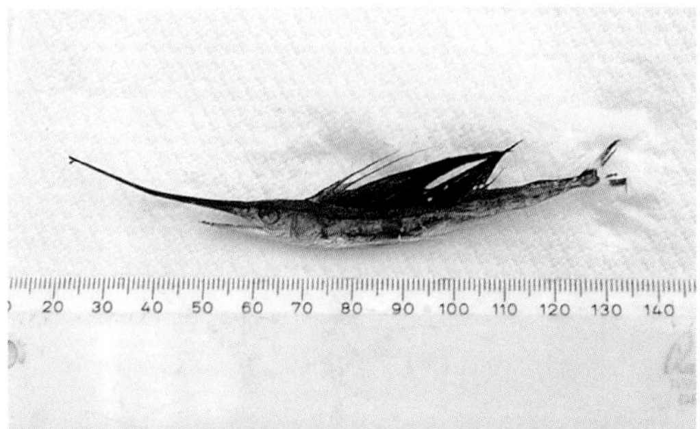

At this size billfish are very difficult to identify.

The shortbill spearfish was fairly quickly eliminated (even though small ones have a high dorsal and a long bill) since all the spearfishes have an anus located well forward of the anal fin (don't ask me why). Our specimen's anus was positioned just in front of the anal fin, so scratch the spearfish.

The next step would have been to take it to the Australian Museum for an X-ray. Black and striped marlin have a slightly different arrangement of their vertebrae, a fact I had resorted to using when I was trying to identify a small billfish from Queensland. However, a colleague, Peter Goadby, jogged my memory about another feature which was potentially useful in separating at least black marlin from the other billfish. Over the years, I had noticed that the two little fins towards the tail (the second dorsal and second anal fins) are arranged differently in black marlin. Measuring many fish indeed showed that, only in black marlin, the second dorsal fin is usually forward of the second anal, while in all the other billfishes, the reverse is the case. I therefore measured the positions of the fins on the specimen in question, and, voila (or, eureka), it was a baby black.

(This could have been the first character to check, but it may not have been conclusive, and in any case, the process of elimination was a very useful exercise, since it led to the same conclusion.)

This fish is important for a number of reasons, not least of which is the fact that it is the first record of a black marlin of this size being taken anywhere in the world. (Some larvae have been sampled, and a couple in the 4 kg range have been caught in Queensland.)

A rarely seen juvenile black marlin (3.9 kg) caught off Townsville. This specimen was actually aged at 140 days old.

The specimen was later studied in more detail to provide some vital information on several aspects of the biology of the species. Radiographs positively identified it as a black marlin. Then, the tiny ear bones (otoliths) were carefully removed. When these were viewed under an electron microscope in Hawaii, they revealed the age of the fish at 80 days, thus helping enormously in determining spawning times, rates of dispersal and most importantly, early growth rates of the species. One of the reasons scientists think that the finding of small billfish is so infrequent, is that they don't stay small for very long, ie. their growth rate is extremely rapid.

After all this poking and prodding, the little billfish will be given a catalogue number, and reside permanently in the Australian Museum, where it will be available for future study.

Quick thinking by several people ensured that this fish find received the attention it deserved. The best thing if you ever come across an unusual catch, is to make sure that the fish is handled carefully, preferably photographed fresh, and then frozen. Contacting the nearest natural history museum or fisheries expert will then determine if the find is indeed valuable.

Do Marlin Change Sex?

Believe it or not, the question of whether or not marlin change sex is probably the most common one I have been asked during my years of mixing with and talking to gamefish anglers. To many people, this might seem an odd query, but with a little background in marine science, it is not only a good question, but a perfectly legitimate basis for some research, both in the field and in the office. There is indeed a perception out there in fishing land that marlin do change sex, and it seems that this is for two reasons: Firstly, it has been proven that all blue and black marlin that weigh over about 180 kg are female, and secondly, some other well-known species of fish, such as barramundi, some wrasses and some flatheads, in fact do change sex. In these cases, all individuals of the species start out their lives as either males or females, changing sex when they reach a certain size (or age). This might seem rather bizarre to us mammals, but for some species of fish, it is a strategy that has evolved so that as many eggs as possible are produced each year, by ensuring that the largest individuals in the population are female. As mentioned, the largest blue and black marlin are also female, but this is due to other factors, as we shall see.

A 350 kg black marlin takes to the air. All black and blue marlin over about 180 kg are female, so do they change sex?

So returning to the question, do marlin change sex? To put it shortly and simply, the answer is a definitive no! We know this from sampling and sexing many hundreds, if not thousands of marlin at game fishing tournaments around the world, and also through the early records of Japanese research longlining cruises. My own experience over the last two decades or so is instructive. In the late 1970s and early 1980s, I was able to examine and sex hundreds of juvenile black marlin from the annual "run" along the Australian east coast (these days, of course, virtually all of these are tagged and released, so it was fortunate that the opportunity was taken back then). This annual influx of black marlin usually ranges in size from 15 kg to about 80 kg, with the great majority averaging about 30 kg. At that size, these fish are reliably estimated to be only one year old, and are sexually immature. Dissecting hundreds of fish over time revealed that the sex ratio of these one-year-olds was very close to 50:50, meaning that unlike barramundi and some flatheads, marlin start out life in equal proportions of males and females. Also, no hermaphroditic gonads of marlin (ie, exhibiting both sexes) have ever been found or reported, in contrast to other species, such as some flatheads, which do change sex.

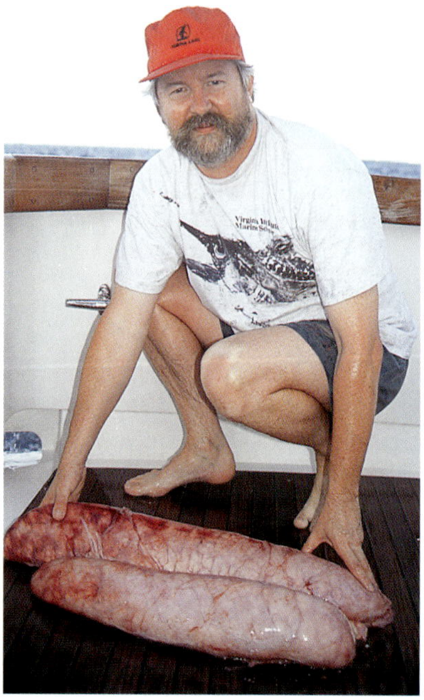

The paired egg sacs (ovaries) of a 450 kg fish may contain over 200 million eggs.

The last piece of evidence comes from studies on age and growth of marlin. Although there is still some doubt about the accurate ageing of marlin across all sizes, these studies strongly suggest that male blue and black marlin grow more slowly than females, and that they probably don't live as long. This fact in itself explains why the largest male black marlin yet recorded weighed 186 kg – one I examined at Port Stephens New South Wales only a couple of years ago. Perhaps surprisingly, the other Istiophorid billfishes (sailfish, striped and white marlin and spearfishes) do not exhibit obvious sexual dimorphism, as do blue and black marlin, since it appears that males and females grow to about the same maximum size in each of these species. In contrast to blue and black marlin, yellowfin tuna and dolphin fish do the opposite: the largest fish are always male. In the case of yellowfin tuna, all fish over about 60 kg will be males. This reverse situation is entirely due to energetics since yellowfin (and other tunas such as bigeye and skipjack) and dolphin fish, spawn hundreds of thousands of eggs every second day for months on end. This means that females of these species are putting most of their energy into egg production rather than growth, while males continue to grow. Blue and black marlin have approached the problem in another way. They spawn up to

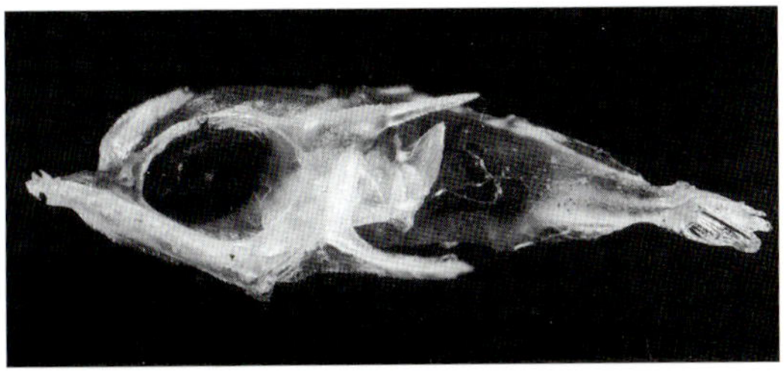

All marlin begin life as tiny larvae, only about 3 mm long, but nevertheless, they are miniature predators of the micro marine world.

200 million eggs possibly several times a year, so natural selection has favoured the development of very large females – in effect, huge, long-lived egg producing machines that can be fertilised by numbers of small, expendable males which "burn out" at a relatively young age. As a female biologist colleague of mine once said so well: "Eggs are expensive but sperm is cheap!"

The other related question is – can the sex of marlin be determined externally, ie, without dissecting it and preferably while it is still alive? The short answer is no, since there are no external features of marlin that are sexually dimorphic. Theoretically, a small piece of muscle from a live marlin could be taken back to the laboratory and the sex determined by DNA fingerprinting, and this may well be possible in the future. But at the moment, the only way to be really sure of the sex of a marlin is to dissect it and examine the paired gonads (sex organs) which run along either side of the intestine. Mind you, it takes quite a bit of practice, and it's amazing how many bets I've settled by determining the sex of a marlin at the dock for the crew. Not surprisingly, at last count, the ratio was still running at about 50:50 male to female.

Mega Yellowfin

One of the largest yellowfin tuna captures happened off the coast of New South Wales at Bermagui. It was taken near Montague Island and weighed in at a massive 124 kg. But what makes its capture interesting was the spate of other big yellowfin that were caught in the same area around the same time. The sizes of the other fish were: 95.5 kg, 98 kg, 102.5 kg and 105 kg. And if those weren't enough, there were two commercially caught fish, both weighing 88 kg gilled and gutted, which would make them 95 kg plus. One of those was also caught at Montague Island (on rod and reel) while the other was long-lined off the continental shelf wide of Bermagui.

Big yellowfin tuna, Thunnus albacares, *like this 103 kg specimen, are becoming rare on the south coast of Australia.*

The 124 kg yellowfin is by far the largest yet caught off the Australian coast, and records have been kept now for over 50 years. The 105 kg fish was the largest prior to that, and the 102.5 kg fish equalled the all tackle record, set off Bermagui in 1984. Of course, several large fish have been landed by local commercial long-liners,

but my investigations have revealed that the largest of those went just on 100 kg gilled and gutted, which would put it at about 110 kg whole. Rummaging around for records of large yellowfin, several people were sure that one of 118 kg (again, gilled and gutted) was weighed at Ulladulla co-op, but a call to the manager there put paid to that rumour – the largest he could recall weighed less than 100 kg.

Checking the weights of fish recorded by Bermagui Big Game Anglers Club for the past 30 years, I was surprised to find that there have been only two earlier yellowfin weighing more than 100 kg, both taken in 1984, and only six others which weighed over 95 kg. These five fish then, three over 100 kg, all caught within weeks of each other, are therefore all the more remarkable.

When the mega yellowfin was caught, local game fishing identity Frank Foster called me and offered to collect scientific information. After a cast was taken, the head was saved for the otoliths (ear bones) and Frank measured the sickles (the second dorsal and anal fins). Earlier studies had pointed to the possibility of two distinct types of yellowfin off the east coast of Australia, most recognisable by the relative lengths of their sickles at the same size. Well, this monster fish had quite short sickles for its size (much shorter than those of similar sized fish

A superb yellowfin tuna about to be released. Note the extended second dorsal and anal fins, which become longer with age and size.

from the central Pacific Ocean), which leads to speculation that it may well have been a resident "shelf" fish which has never been far offshore. This may also be the case for the other big fish mentioned, since photographs of two of them also indicated short sickles.

The otoliths of this fish, measuring about a centimetre long, were duly retrieved by CSIRO scientists and probed and prodded to reveal their secrets. They were sectioned, the banding patterns enhanced, put under the gaze of powerful microscopes and subjected to computer enhancement. After this painstaking treatment, scientists were able to come up with a confident, if qualified, estimate of the age of the fish. They put it at 13 years old, give or take a year, but made the important point that this was only true if each ring or band in the otolith represented one year of growth.

At this stage, the one band equals one year assumption is reasonable, based on previous work on ageing of yellowfin, so this fish was, in all likelihood, only about 13 years old. This surprised me a little, since we were confident that fish of 70 to 80 kg were about 7 or 8 years old.

Above: A catch of school yellowfin tuna. The small fish would be in their first year of life, while the bigger fish on the left would be two to three years old.

Right: Large yellowfin tuna are a prized catch

Fish grow throughout their lives, but as they approach "old age", the growth rate slows down. This means that very large fish (for the species) have usually been growing very slowly in the latter stages, so they are usually quite old. However, in the case of this unusually large yellowfin tuna, it seems that its growth rate had been quite rapid throughout its life. Perhaps a classic piscatorial case of the adage "live fast and die young".

Tiny Tuna

There's something about baby animals that brings out a certain awe and sense of wonder in we humans. I don't just mean the parental cluckings over cuddly baby lambs, rabbits and puppies, but rather that general fascination in seeing miniature offspring of larger, familiar critters – and here, of course, I'm talking baby fish.

All pelagic fishes, billfish included, produce huge quantities of eggs so, in theory, their offspring must be extremely common. This may well be so, but it is surprising how rarely very young billfishes and other pelagics have been caught and recorded by science. However, every now and then, a baby marlin, sailfish or broadbill swordfish is caught and, knowing my interest, cooperating anglers and skippers have forwarded the fish to me. Such specimens are very valuable for carrying out studies of comparative growth rates.

My success in obtaining tiny billfishes from cooperative anglers over the years has not been matched by a similar influx of baby tunas (or other pelagics for that matter). Tiny tunas are apparently non-existent in Australian waters, or at least, in Australian anglers' catches, but this is not necessarily surprising.

A row of tiny skipjack tuna, Katsuwonus pelamis, *caught by bait jigging in Papua New Guinea. These fish are probably no more than a month old.*

I finally had the chance to see some very small tuna up close and personal for the first time in the Huon Gulf off the north coast of Papua New Guinea and the port of Lae. Well-known angler John Cross was my host, and prior to my arrival, he had excited my interest with stories of baby tuna commonly caught in the area. This is indeed a fascinating area for tropical pelagics, being very deep close to shore, and fed with a constant supply of logs and trees from many huge rivers. The logs and other flotsam provide a vast array of habitat for small bait-fish, as well as for juveniles of many species, including rainbow runners, various trevallies and at least two species of tuna.

One problem here, though, is that there is so much natural floating debris that small pelagics can sometimes be difficult to locate since they have so many places to hide. This has resulted in the Lae Game Fishing Club putting down several impressive FADs (large, permanently anchored fish-attracting buoys) to give nature a helping hand, as it were. Perhaps because they are relatively fixed in space, the FADs tend to consistently out-fish the moving logs, possibly because the fish use them as consistent reference points. Whatever the reason, the FADs work, and this is where the tiny tuna can usually be found.

The most common baby tuna appeared to be skipjack or striped tuna. They were present in all sizes, the smallest I saw being only about 4 cm long. I say, "appeared to be skipjack", since they did not have the belly stripes of adults, but did possess all of the other characteristics of the species.

The other species of tiny tuna commonly caught off Lae, pictured right, are most likely baby yellowfin. I say, most likely, since it is also possible that they could be bigeye tuna – at small sizes, they are very difficult to separate from yellowfin. However, the rather rounded tip of the pectoral fin in this case point to it being a yellowfin.

The biology of the tropical tunas is fairly well known, and explains quite a bit about my observations. Both skipjack and yellowfin are tropical spawners, hence the lack of tiny ones off most of the Australian east coast. Skipjack are known to spawn across vast areas of the tropical western Pacific Ocean, while yellowfin also have broad spawning areas through the tropics. Not only are the spawning areas of these species vast, but their extraordinary reproductive biology also ensures a constant

Dogtooth tuna
Gymnosarda unicolor
150cm
130 kg

Actually a giant bonito, the dogtooth has a fragmented tropical distribution around islands, reefs and offshore atolls. Juveniles are not well known, and may have a more pelagic habitat than adults, which rarely move far from home reefs.

A beautiful miniature tuna from the Huon Gulf, Papua New Guinea. Is it a yellowfin or a bigeye?

supply of teeming hordes of newly hatched tuna larvae. The average adult female skipjack tuna spawns hundreds of thousands of eggs every day for months on end, while yellowfin tuna are apparently only slightly less amorous, spawning only every second day. Considering that most species of fish have short spawning seasons, and spawn only once a year, this strategy of saturating the ocean with youngsters has obviously been very successful.

This frenetic reproductive activity (as well as very rapid growth rates) has a very fortunate benefit for the tunas, especially skipjack. With such an enormous rate of biomass turnover, these so-called "cockroaches" of the ocean have so far been able to keep one step ahead of the ever-increasing fishing power of the western Pacific tuna fleets. More than a million tonnes of skipjack and hundreds of thousands of tonnes of juvenile yellowfin are purse seined and poled throughout their range every year, yet all of the scientific models and analyses (and there have been plenty) agree that the stocks are still under-fished. This is good news for these particular species, but let's just hope that their amazing biology can continue to cope with the world's seemingly insatiable appetite for tuna salad and sandwiches.

All in the Genes

I've just packed and sent off a collection of frozen marlin bits and pieces which may well change the way we think about their stocks in the Pacific Ocean. All the samples are packaged, labeled and supplied with a generous amount of dry ice for their 40-hour journey all the way to Virginia, on the east coast of the United States.

I have been busy collecting small pieces of heart, liver and muscle from striped marlin and blue marlin, all caught by game fishermen off the Australian east coast. Their destination is the Virginia Institute of Marine Science, where my colleague, and leading fish geneticist, Dr John Graves, eagerly awaits their arrival. John is trying to unlock some of the mysteries of the population genetics of tuna and billfish. Using a fairly new technique called mitochondrial DNA analysis, he is comparing the genetic makeup of fish collected from widely separated areas to determine if there is more than one stock within their widespread distributions. The main project I am helping John with is a study of striped marlin throughout the Pacific.

Japanese long-lining catch records have shown that the striped marlin distributed around the Pacific Ocean are in a rather odd horseshoe-shaped pattern. Analysis of their size ranges suggest that there may be slow movement of fish as they grow, from the northwestern Pacific across to the east, then southwards along the Californian coast, and then across to the southwestern Pacific Ocean, in particular, New Zealand and eastern Australia. The latter region has the largest

Striped marlin
Tetrapturus audax
260 cm
220 kg

Subtropical to temperate pelagic species occurring through the Indo-Pacific, but in an odd, horseshoe shaped pattern in the Pacific. Tagging and genetic results suggest that there are distinct populations on either side of the Pacific.

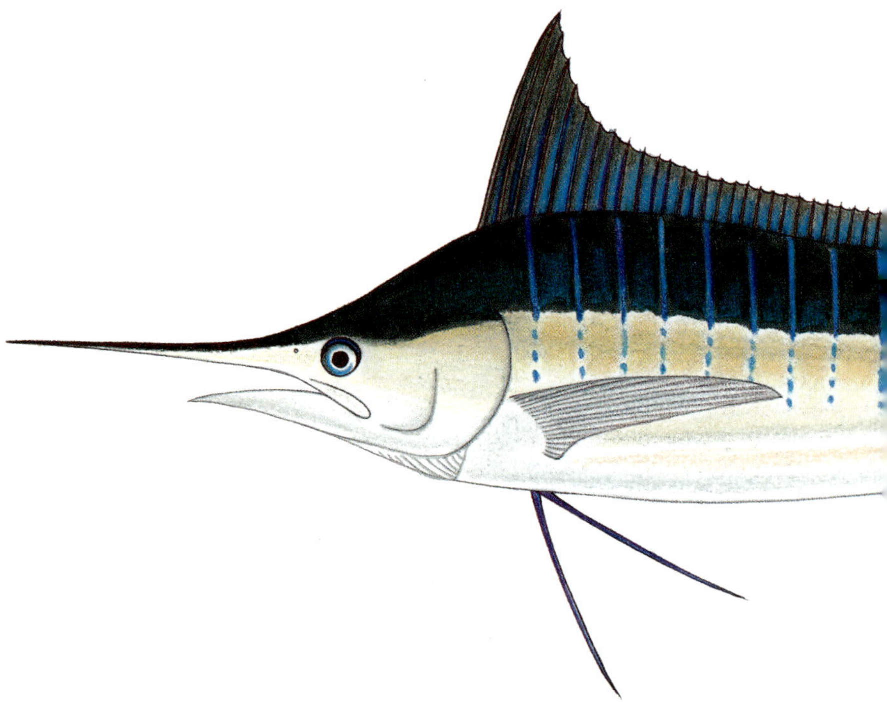

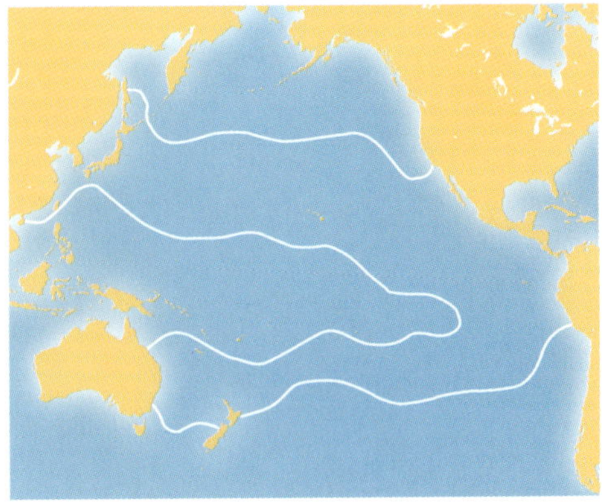

The odd, horseshoe-shaped distribution pattern of striped marlin in the Pacific Ocean.

striped marlin, and spawning takes place in this part of their range.

With limited information, scientists have been attempting to assess the status of the stock of striped marlin (and other species of tuna and billfish) in the whole of the Pacific Ocean, that is, the population size and effect of fishing on that population. The trouble is that good stock assessment requires accurate knowledge of the boundaries of the stock with which you are dealing. If all of the striped marlin throughout the Pacific Ocean are part of one continuous, mingling stock, then stock assessment and consequent management plans would be quite different from those based on the existence of a number of quite separate stocks. Three methods are used to attempt to answer this question.

Tagging is the most direct way of revealing movements of individuals from one part of the species' range to another, and this is where recreational tagging of billfish and tuna has been particularly useful. Over 12,000 striped marlin have been tagged by anglers off Baja, California, and recaptures to date indicate some movement north and south, as well as offshore as far as Hawaii. However, of all of the striped marlin

tagged over there, very few have been recaptured in the Southern Hemisphere, and none in the western Pacific Ocean. Similarly, on the Australian tagging program, over 9000 striped marlin have been tagged, and 70 have been recaptured. None of those had moved more than 1000 km.

The author collecting samples of sailfish from a commercial processing plant in Papua New Guinea.

Morphometrics (measuring proportions of body parts such as fin lengths in relation to head length, etc) can also be a useful tool because variations in such ratios can reflect genetic differences between areas, in effect, revealing "races" of the same species which could interbreed, but don't, simply because they rarely, if ever, mix.

Population genetics can support tagging and morphometrics by determining if the genes of fish from different regions are uniformly mixed, or different enough to indicate separate groups, or stocks. A number of methods have been applied with varying success, but now, the new more sensitive method, mitochondrial DNA analysis, is revealing very important information. Without getting too technical, this method examines the genetic material (DNA) inside mitochondria, which are little energy producing organelles inside most cells, especially in the heart and liver. The mitochondrial DNA doesn't take part in genetic exchange, but is locked up inside the mitochondria, only changing by mutation. This DNA then, represents ancestral genes, and therefore, by comparing its makeup in widely separated groups of the same species (in this case, striped marlin from both sides of the Pacific Ocean) we can tell if the two groups are genetically distinct, and if so, even estimate the length of time they have been separated.

The results of the study on striped marlin is proving to be very interesting. So far, it looks like the striped marlin of the Pacific Ocean may in fact consist of several distinct groups, or stocks, contrary to the earlier theory of one continuous stock. On the other hand, the Pacific blue marlin so far sampled are relatively uniform genetically, while Atlantic blues are more of a mixed bunch.

It is worth mentioning that the striped marlin study is funded by The Billfish Foundation, a private organisation formed in Florida in the mid 1980s with one goal: the preservation of billfish. A case of anglers setting the agenda and getting on with the job of information collection.

Sail Away

I have been very close to the Australian Game Fish Tagging Program, begun in 1974, for a long time, and try to keep in touch with its developments and results. I was therefore very interested to learn of one of the more important recaptures of a tagged game fish in the program's history – a sailfish. But before going into the details of why the recapture was so noteworthy, some background information is necessary.

Over the years, sailfish have been tagged in numbers off several discrete areas on both the east and west Australian coasts. The main areas have been Cape Moreton (south Queensland), Dunk Island, Cape Bowling Green (both north Queensland), Broome (northwest Australia) and Dampier (in the west). Recaptures have been slow but steady, gradually revealing some unexpected facts. It wasn't until 1980 that the first sailfish was recaptured, and since then, 20 sailfish tagged off Cape Bowling Green have been recaptured, 11 from Cape Moreton, nine from Dunk Island, and three each from Dampier and Broome. Now the really interesting point is that all but two of those 46 recaptured sailfish were taken in the exact locations where they were released. Sure, some of them were only at liberty for relatively short periods, some as short as two or three days, but many had been free for lengthy periods ranging from months to years. For example, four of these fish were recaptured near their release points after more than three years, and one after four years. One particular fish was tagged on two separate occasions a year apart before being recaptured the following year.

Another sailfish, Istiphorus platypterus, *about to be tagged. Will it stay or will it go?*

The two fish that moved any distance at all were both tagged near Dunk Island. One was recaptured about 80 km to the north after three years two months, while the other was recaptured 150 km to the south (at Cape Bowling Green) after two years one month.

Now, for the odd fish out: this particular sailfish was first tagged in the Whitsunday Passage (only a few

The sailfish is capable of high speed swimming, but unlike marlin, prefers to stay close to home.

A sailfish about to be tagged and released.

have been tagged in that area) and recaptured four years and five months after its first capture, 95 km wide of Cairns, 500 km to the north by a local tuna long-liner.

Several features make this an interesting recapture. It is the furthest distance travelled by a sailfish for the program. It set a record for time at liberty for an Australian tagged sailfish and the fish was estimated to have grown from about 30 kg to 40 kg during this time. Lastly, it was the first Australian tagged sailfish to be recaptured away from the coast, and by a commercial fishing vessel.

The gorgeous colours of the Indo-Pacific sailfish. Built for speed, but apparently not for long migrations.

Reviewing this recapture in the light of the previous findings, and some other biological studies, can tell us a fair bit about our coastal sailfish populations. Before this recapture, it had been generally accepted that adult sailfish populations tended to remain pretty well discrete. Fish did apparently depart from the various fishing grounds for at least some part of the year (perhaps offshore?), but the same fish appeared to return to the same grounds.

Peter Speare of the Australian Institute of Marine Science conducted a study of parasites found on sailfish in northern and southern Queensland, and found enough differences to support the separate population hypothesis suggested by the tag data. Now this one recapture throws a spanner in the works to some extent, although as we all know, one swallow doesn't make a summer (or some such marine equivalent). You see, genetically speaking, it only takes an occasional individual fish to move between two seemingly separate populations, and to spawn, in order to mix the genes of the two groups past the point of being able to detect differences. What's more, the rate of movement only needs to be, say, one fish per generation to achieve this result.

What is really needed now is a genetic study of east coast sailfish collected from north and south Queensland. If genetic differences are found, then the evidence for separate populations would be hard to refute. Trouble is, acceptance of the tagging program has been so successful that it is now very difficult to convince anglers to bring in sailfish for scientific purposes. Over in the Wild West, I had just such a problem,

A beautiful sailfish about to be tagged and released. Tagging is the only direct way of studying fish movements.

when I needed a handfull of fish to be brought back to shore for a genetic study. For the first three days of my visit to a game fishing tournament at Broome, over 500 sailfish were tagged, but none were brought in for sampling. Finally, with much pleading, I was finally able to convince some of the anglers to bring in small numbers of sailfish for the study. I could fully appreciate their dilemma, and it clearly demonstrated their dedication to tag and release, and to ethical fishing, but I was certainly relieved to finally obtain the valuable samples. After all, biological research sometimes does entail the sampling of a small proportion of the population in order to ensure future preservation of the species. The results of this particular exercise, by the way, suggested that there are no strong differences between sailfish populations found off Broome and Dampier. These two ports are 600 km (320 nautical miles) apart, and while no tagged sailfish had apparently moved between these areas, that does not mean that it never happens, nor that eggs, larvae or small juveniles do not make the trip from time to time.

Relating the story of a single recapture of a tagged sailfish helps to demonstrate the value of recreational tagging programs. Every year, tens of thousands of fish are tagged and released by an army of interested anglers as a purely voluntary exercise. In doing so, they take part in a large-scale scientific experiment which can and does constantly reveal new information about the great fishes of the open ocean. And that's something about which they can feel justifiably proud.

Twice Bitten, not Shy

Each year, thousands of marine game fish are tagged and released by recreational anglers, who do so voluntarily, and return all release information to the New South Wales Fisheries Research Institute on prepaid postcards. It's all part of the Australian Game Fish Tagging Program. Every now and then, the program produces a surprise or two.

A yellowfin tuna recaptured off the Australian southeast coast was duly reported to Fisheries; but it was then that the plot thickened. Matching the records revealed that this fish had in fact been previously tagged not once, but twice. The story began when Christian Curwen-Walker of Victoria tagged a small yellowfin, estimated at about 8 kg, wide of Bermagui, one of many tagged at the time. The fish was then caught and tagged again some 52 days later in the same general vicinity by Ken Stockton, also of Victoria. The interesting aspect here was that Ken didn't notice the first tag, and so now two tag cards for the same fish resided in the files of NSW Fisheries. The story was finalised when, 662 days after the first release, Laurie Chadder, fishing off Sydney, boated the fish, now about 30 kg, and spotted both tags.

Double recaptures of tagged fish are rare, and they give us more information than the usual single recapture, which show that a fish had swum from point A to point B. Adding a point C can say a lot about the overall movements of a fish. A lot of earlier recaptures of tagged yellowfin clearly demonstrated that juveniles stay associated with

Yellowfin tuna
Thunnus albacares
210 cm
150 kg

Tropical to temperate worldwide tuna species, occurring right around Australia except for southern ocean.

Tagging a black marlin off Cairns. Multiple recaptures are rare, but can and do happen.

the continental shelf of southeast Australia for at least two years, moving slowly northward in winter and southward in summer. The double recapture provided some extra information about the length of time the fish had remained near its point of release before heading off on its annual cyclic trek. This is the first case of a yellowfin tuna knowingly being tagged and recaptured twice, but there have been other notable multiple recaptures for other species.

SAILFISH

Perhaps the best known double recapture of a tagged game fish was a sailfish which was landed near Dunk Island. It wasn't until the fish was weighed that not one, but two tags were noticed in the fish (tags often become covered with marine growth) and after checking details, the story emerged that this fish, like the yellowfin tuna, had been hooked, tagged and released on two previous occasions. The original tag going unnoticed the second time. It was first released in the same area according to the records more than three years before. Two years later, again in the same area, the fish was recaptured and tagged, swimming free again for another year before its final capture. This particular recapture has been important in theorising about localised sailfish populations, and in helping us realise that many tagged fish are probably hooked and re-tagged without the original tag being noticed. In areas where tag-and-release is the norm, not noticing previously placed tags may help to explain apparent low recapture rates.

YELLOWTAIL KINGFISH

The species which has shown the highest number of multiple recaptures is the yellowtail kingfish, and many have been tagged and released twice, often over fairly short time periods, and nearly always in the same location. This is partly explained by the tendency for juvenile kingfish to stay in the one area, resulting in high recapture rates for the species. However, one particular fish set a record some time ago which has proven hard to beat. This particular little kingie was first tagged at a wave recorder buoy off Coffs Harbour, and subsequently recaptured and released by different anglers at the same site no less than seven times. The fish became a bit of a celebrity and I recall following its adventures with great interest at the time. Maybe the fish had suicidal tendencies, or maybe it was just inevitable, but the seventh time it was recaptured happened to be its last when it was deep hooked and kept for dinner. The periods between tagging and recapture were 4, 44, 4, 7, 12, 45 and 31 days, so this piscatorial Houdini had been hooked, handled (so that the tag could be read) and released on six separate occasions, and hooked one more time, all in the space of 21 weeks.

Encounters like this and the other examples mentioned, certainly provide a lot of food for thought about survival of tagged fish. They also indicate the need to be on the lookout for tags in any fish which you catch, especially if your intention is to release it; and don't forget, by the way, to check out both sides of the fish for tags prior to release.

All of this of course leads to the question: if you catch a tagged fish, should you keep it, or should you let it go again? There are arguments for both cases. Keeping the fish and freezing it will allow the scientists involved in the tagging program to examine it and take samples for growth studies. On the other hand, multiple recaptures, even though they are rare, do yield a lot of information. My rule of thumb would be to re-release the tagged fish if the tag looks fresh, or if the fish is small enough to measure accurately (record the original tag number rather than remove the tag and place another, by the way). If the tag looks old, then it is probably better to keep the fish and contact the address or phone number on the tag as soon as possible. A tricky call, but worth the time and effort either way.

A yellowtail kingfish, Seriola lalandi, *holds the record for multiple recaptures, – six, at last count.*

Fascinating Fish Facts

From the perspective of air breathers like us, fish can seem to be almost alien life forms. Sure they are fellow vertebrates, but they breathe water after all, and live in a world which is incomprehensible to most – even though water is our ancestral home. But science shows us how fascinating fish can be; how they have adapted so perfectly to their environment, often in quite remarkable ways. The life stories of fish and other denizens of the deep are indeed fascinating, and definitely worth the telling.

The world of some fish is filled with colour

And Fish Might Fly

It always thrills me when a school of flying fish showers into the air and glides effortlessly over the wave tops, dropping back with a splash, one by one. Sometimes large ones, sometimes tiny, but always beautifully coloured in silvers, blues and purples, and always fascinating. When you think about it, a fish that can fly is a pretty amazing thing, but fly they do, or at least glide in an efficient and effective way.

So how and why do fish fly? To begin with, flying of sorts has evolved in several separate groups of fish. The well-known oceanic flying fishes are the best examples. I'll get to those shortly, but it may surprise you to learn that there is also an African freshwater flying fish which catches insects in mid air, as well as the South American hatchet fish which has large pectoral fin muscles. It is the only fish that can fly short distances by actually flapping its wings – er – fins.

Another fish for which flight has been suggested, but not necessarily proven, is the flying gurnard. One of my favourite-named fishes, the purple flying gurnard possesses

Flying fish
Exocetus sp
30 cm
1 kg

Several species of flying fishes occur in Australian waters, most being tropical to sub-tropical in their distribution. There are two-"winged" and four-"winged" varieties, the latter having expanded pelvic fins. The one illustrated here is a four-winged species.

As flying fish take to the air, their tails beat rapidly, leaving telltale patterns along the watery runway.

enormous pectoral fins that do look like wings, but which are not built for flight. Consulting a wide range of fish books reveals that some authors repeat the idea that these gurnards can make short, rather inefficient flights. However, such statements seem to be made from a theoretical basis, and I am yet to find a first-hand account of someone actually observing gurnards flying.

The fabulous flying fishes of the open ocean, which certainly do fly, belong to a single family containing about 40 species. All are torpedo shaped, silvery fish, and it is generally accepted that this large group of flying fishes evolved from the garfish family (the so-called halfbeaks or ballyhoo). In fact, a couple of the Atlantic garfishes make long leaps from the water with their fins spread – an obvious precursor to the true gliding flight of the more advanced fliers.

Flying fish are commonly observed, and for many years there was hot debate among scientists as to whether they flapped their pectoral fins in flight or merely glided. This problem was not solved until the early 1940s when film of flying fish shot at night with strobe lighting was carefully analysed. This clearly showed that the pectoral fins are used only for gliding, and that becoming airborne was achieved by a leap, followed by very rapid beats of the tail (50 to 70 strokes per second). The tail has a very long lower lobe to maintain propulsion while the body of the fish is leaving the surface of the water.

More observations and analysis of flying fishes since then have revealed some pretty amazing facts. Adult flying fish of various species range in size from 15 to 50 cm. There are two-winged and four-winged flying fish, the latter growing to the largest sizes, and making the longest flights. Take-off speed is probably about 35 km/hr, reaching airspeeds of up to 70 km/hr. The longest flights recorded are about 400 metres, but these are mostly achieved by dipping the tail back into the water a couple of times during the flight for an extra burst of propulsion. It is also thought that the "ground effect", that is, pressure between the wings and the ocean surface, may help the flight (as it certainly does the effortless ocean-skimming of seabirds such as shearwaters). In rough weather, flying fish have been observed to reach heights of at least 10 metres, and I vividly recall an uncomfortable day off Bermagui when I looked up at the outrigger to be startled by a couple of foot-long fish soaring overhead like a pair of gliding mullet.

Flying fish only occur in warm waters (above about 20°C), and it has been calculated that they probably can't summon the energy required to achieve flight at lower temperatures.

Scientists are still uncertain whether flying fish have control over the direction of their flight or not. They don't seem to manoeuvre much in mid air, but their eyes look downwards and can almost certainly focus in both air and water. This is also borne out by the fact that they often seem to land on purpose in weed lines (good sources of food), so it is likely that they can control their flight to some extent. Then again, it is not uncommon for flying fish to land on the decks of boats in broad daylight, so flight control may not include avoiding large, unidentified floating objects.

An 18th-century illustration of two-winged and four-winged flying fishes. It was not until the 20th century that the nature of their flight was discovered.

Finally, the best answer to the big question, "why do flying fish fly?" is probably simply to escape predators. It's certainly a good tactic to disappear from view when being chased by

The remarkable flying gurnards, Dactyloptera spp., *are still thought by some to be capable of flight. However, it is now generally accepted that their "wings" are used for scaring off predators.*

a marauding marlin or a hungry dolphin fish, however seabirds are known to pick up fish in mid flight, and making a big splash on re-entry can often alert a lurking predator. Nevertheless, the trade-off must be on the positive side, since flying fishes are certainly prolific in the world's oceans.

One other point: many species of flying fishes belong to the genus *Exocetus* – the name rather appropriately adopted by the French for their famous ocean-skimming Exocet missile. It's a bit of trivia that always comes to mind when I'm lucky enough to see a little, living "exocet" gliding gracefully across the briny.

Smell and Taste in Fish

I strongly suspect that many a patient angler often tries to imagine what it might be like to actually be a fish. What does the quarry see, how sensitive is its sense of smell, and having found the baited hook, does its actual taste really matter? Fortunately, for a whole range of reasons, these questions are also of interest to scientists. Casual observation of most fishes will reveal that they must have reasonably good eyesight: after all, the majority have well-developed eyes. But water clarity in many habitats is often very poor, which is where the senses of smell and taste must come into play in a big way. Smell and taste are, of course, quite different, but they are related in that both are geared to detect substances dissolved in the water. Smell is used to detect something (usually food) in the short to medium distance – to astonishing effect sometimes – while the sense of taste requires actual physical contact.

Smell

Look at any fish and you will see evidence that they can smell. Their nostrils, or "nares" to be precise, are usually just in front of their eyes (in sharks, on the lower surface of the head) and in many cases are paired. Unlike other vertebrates, however, their olfactory organs are not used for breathing, but are usually deep pits lined with highly convoluted membranes. The convolutions increase the surface area of the organ, thereby greatly increasing its sensitivity. Some fishes, such as the piranhas,

Eeltailed catfish
Tandanus tandanus
90 cm
7 kg

Freshwater species occurring throughout the Murray Darling river system of southeast Australia, and introduced to many eastern flowing rivers. Like all catfishes, the eeltail uses its highly sensory barbels to taste its environment and find food.

have greatly enlarged nostrils, no doubt to detect blood at great distances, while others such as members of the puffer fish family have their nasal openings at the end of short stalks. Sometimes, sense of smell is important at different stages of the life cycle but not others. In the case of the bizarre angler fishes, the ones which lie in wait and wiggle a bait in front of their mouths, smell is not very important to this method of feeding, and so their nasal cavities are small. However, the juvenile males of some species of anglers have huge nostrils by means of which they find a much larger female, presumably via pheromones, whereupon they attach themselves firmly to her, becoming fertilising parasites for life.

Just how sensitive the sense of smell is in some species of fish is perhaps best demonstrated by Pacific salmon. For many years, researchers knew that salmon nearly always returned to their natal streams to spawn after spending years offshore, but the mechanism was hard to pin down. Intensive studies have eventually shown that navigation towards the coast is achieved by orientation to the sun, and perhaps the earth's magnetic field, but once near shore, the sense of smell takes over. Each stream has a unique cocktail of natural odours and the salmon home in on the sweet smells of their childhood. The really amazing fact here, though, is the incredibly small concentrations of odour the salmon must be able to detect. In fact, as one of my favourite authors, Brad Matsen, put it so neatly in a recent book of his, in finding

their stream of origin, salmon can detect the equivalent of one drop of vermouth in 500,000 barrels of gin. A very dry martini indeed.

Then, of course, there is the case of the fabled catfish of some South American rivers, *Vandellia cirrhosa*. Attracted from great distances by the smell of urine (why, we can only guess), this 5 cm fish may actually swim up the urogenital openings of human waders or swimmers. As if that wasn't bad enough, finding itself in such a predicament, it then extends its spines and firmly wedges itself inside, causing great pain, only being able to be extracted by surgery. Makes one wince at the thought. I used to think that this story was a myth, but have found reliable references attesting to its truth. In the absence of humans, these fishes are actually parasitic on other fishes, and probably use their sensitivity to excreted nitrates to find their host. They then enter orifices such as gill openings and nip off morsels of food.

The goatfishes have well-developed tasting barbels under the chin. Here, a southern goatfish, Upeneichthys vlamingii, *"tastes" for a feed.*

TASTE

Fishes have taste buds similar to ours, and most have such sensory cells in their mouth, or on their tongues, if they in fact possess one. That's easy to relate to, but in the weird world of fish, taste buds may also be found on the outside of the body. Carps, cods, mullets and sturgeons all can "taste" over much of their body surface (imagine being able to taste anything you picked up), while many species have evolved long tendrils or barbels near their mouths or under their chins that are covered with taste buds. The latter would include catfishes, goatfishes and many deep-sea fishes, some of which look like they have grown huge, trailing, branched beards. In all these cases, the fish live in turbid or deep, dark water, and eyesight has become unimportant in finding food.

The sense of taste is obviously highly developed in most fishes (although I can't vouch for tiger sharks, which seem to swallow just about anything), and the thinking angler will always benefit by spending some time observing fish feeding in an aquarium. It will become pretty obvious that most fish will mouth any likely looking item, but will very quickly eject it if it doesn't come up to its gourmet expectations. So the message from all of this is pretty clear: fresh, natural bait and berley are a must. But if you are an angler, you already knew that, didn't you.

Built for Speed

Because we humans are creatures of the land and air, the idea of moving through water at high speed is not a natural concept to grasp. The top Olympic swimmers may negotiate the 100 metre freestyle at about 7.5 km/hr (with a flying start and a good shove off the wall at the halfway mark). However, that sort of speed is snail's pace to a wahoo or sailfish, both of which have been clocked at close to 100 km/hr – at which speed they would complete the 100-metre dash in under 4 seconds.

High-speed predators are certainly superb examples of evolution of body form towards a purpose, in this case sheer speed. Because water offers so much resistance to movement, the predators (and some of their prey) have all evolved an extremely streamlined shape together with other features that all reduce drag. In most of the mackerels, tunas and billfishes, the fins fold beautifully into slots for high-speed bursts. Even the huge dorsal fin of the sailfish folds perfectly into a long groove along the back. I am often asked about the function of the bill of billfish, and while it can certainly be used as a slashing or striking weapon, some scientists think that the bill's main role is to maximise streamlining, piercing the water first in the manner of the spike on the nose of a Polaris missile. All of these fishes also have other drag-reducing features such as lateral keels on the tail wrist and single or multiple finlets in front of the tail. These adaptations no doubt help to reduce turbulence at the body surface,

Albacore
Thunnus alalunga
100 cm
22 kg

A tuna which prefers cooler, temperate surface waters, found around the southern half of Australia, and across the Pacific and Indian oceans to about 45° south. Also occurs in northern Pacific. A typical, streamlined tuna showing amazing adaptations for high-speed swimming.

but as far as I am aware, their actual functions have not been studied under laboratory conditions. Submarine design no doubt owes a fair bit to copying nature's solutions to reducing friction, and I sometimes think that a submarine in the shape of a blue marlin might be a winner (as well as being aesthetically pleasing).

Measuring speed

Large fish are difficult to study in captivity. The speeds of sailfish and wahoo mentioned above were estimated by measuring the rate of line peeling off a high-tech reel when the fish were hooked. However, quite a bit of work on burst and continuous swimming speeds has been done on a variety of captive fish.

Many of these studies are summarised in a paper by J Videler and CS Wardle (*Reviews in Fish Biology and Fisheries, Volume 1*, pages 23-40, 1991). These authors reported on a wide array of fish species swimming in circular tanks and analysed for such factors as tailbeat amplitude, maximum and sustained speed, and so on. Scanning their results show some interesting consistencies. For comparative purposes, speeds are measured in terms of body lengths per second (L/sec), and maximum speeds measured in this way include the following: goldfish (3.5-9.3 L/sec), rainbow trout (1.5-11.6 L/sec), Atlantic cod (2.8-3.5 L/sec), slimy mackerel (5.3-18.0 L/sec) and bluefin tuna (9.3 L/sec). While, in general, smaller individuals of a given species will have faster maximum speeds in terms of body length per second, the interesting finding is that the maximum relative speeds are pretty comparable. Having said that, when the body lengths of the bluefin tuna (2.3 metres) and goldfish (7 cm) are taken into account, the absolute maximum speeds measured were 77 km/hr for the tuna and 2.3 km/hr for the goldfish. Sustained speeds have also been measured for a variety of fish, and these tend to range between about 1.5 and 4.0 L/sec (average about 2.0 L/sec) for a wide variety of species. Again, this is a narrow range but, of course, large fish will cover much greater distances than small fish over the same time periods.

The propulsion ends of yellowfin (top) and bigeye (bottom) tunas. Note the narrow tail wrists and the rows of finlets for reducing friction.

The superb streamlined shape of the fastest shark in the ocean, the mako, Isurus oxyrinchus.

I've been interested for quite a while in the maximum rates of travel of tagged game fish recorded on cooperative game fish tagging programs. For species such as tuna, billfish and mako sharks, the maximum rates of movement, as measured in kilometres travelled over days between release and recapture, often average about 20 to 30 km/day for fish of about 2 to 3 metres in length. This works out at an average sustained swimming speed of only about 0.1 body lengths per second over periods of up to a year or more. When I was involved in tracking black marlin using sonic tags, they tended to swim along at a leisurely 2 knots or so. This translates to about 0.3 body lengths per second, so the overall averages derived from tagging programs seem to be in the right sort of range for sustained swimming over long distances.

Obviously, extreme speed is only used for hunting or fleeing, but when they do turn on the power, the pelagic predators prove that they really are built for speed.

Mucus and Slime!

Not a very appetising title, I know, but then again, you're not a fish. If you were (and if you could read), then you would be vitally interested in mucus and slime – because without it, you wouldn't last long in or out of the watery world.

As anyone who has caught, handled or prepared fish for the table would be aware, all fish are covered with at least a thin layer of mucus. The sometimes amusing common names of some fish graphically emphasise this aspect. Slimy mackerel, slimy flathead, slimeheads (the orange roughy is one), slippery, slippery dick, soapy, butterfish, and my favourite, the snotnosed or snotgall trevalla (or just plain snotty) are names that effectively convey the image of slippery, slimy denizens of the deep. (By the way, for fairly obvious reasons, the "snotty" is marketed under the more exotic name "warehou".)

The well-named slimy mackerel, Scomber australasicus. *All fish are covered with a thin film of slime which may be distasteful to some, but is vital to a fish's existence in the watery world.*

The slippery feel of fish results from the high water content of slime together with large gel-forming molecules, or glycoproteins, also known as mucins. These large molecules are present in the mucus and slime of most vertebrates (including us), and all are able to swell in water, and to entangle, giving the water a semi-solid structure.

Obviously mucus and slime must have important functions in fish, otherwise why would they all be covered in the stuff? Scientists have indeed pondered this very question, and have come up with a whole host of functions that might be attributed to fish mucus. These include protection from abrasion, defence against predators (physical as well as toxic defence), disease resistance, feeding, nest building, lowering resistance for swimming, balance of body salts, and so on. All of these functions probably do occur in various species of fish, but let's consider the ones considered the most common.

Protection

General protection seems to be the main function of slime, on both the micro and macro levels. Slime forms a non-stick barrier to invading bacteria, and as it slowly wears away and is replaced, it takes harmful organisms with it. This is the main reason that anglers are always encouraged when handling fish intended for release to minimise contact with hands, especially dry hands. Tagging of small fish involves some handling, but plenty of recaptures of tagged fish show that recovery from handling certainly occurs. However, too much loss of body slime could spell disaster for a fish, so if you must remove a fish from the water, try to do so with wet hands or a cloth.

Another example of the importance of protection by slime from disease is shown by the phenomenon of red spot disease in estuarine fishes on the Australian east coast and throughout South East Asia. This problem, also known as Bundaburg disease, shows up as ugly red ulcers along the flanks of species such as bream and mullet in many coastal rivers, often causing major problems for commercial fishermen. Although there have been plenty of theories about the cause, it wasn't until recently that scientists from New South Wales Fisheries cracked the problem. Coastal rivers often become quite acidic in times of drought due to historic land draining and exposure of acid sulphate soils. The acidic water can erode the thin slime layer of fish that are already stressed from this exposure. Then, invading bacteria and fungi can gain a foothold inside the skin of the fish, eventually causing the unsightly sores and ultimate death of the unfortunate victim. It is even known that fish slime probably has a fungus-inhibiting factor in its chemical make-up, again making it all the more important for protection. And finally, fish slime may even contain substances promoting the healing of wounds, so the message is, handle that fish carefully if you intend to return it to the water.

The young of some species of fish, such as freshwater Cichlids, follow their parents and feed off their body slime. It seems that as well as providing nourishment, the slime transfers immunity to disease in the form of antibodies contained in the mucus.

Body slime can also protect against predators, not only by making the fish hard to grasp, but also by being foul-tasting or even toxic in some species. Certainly, some species of fish, like the primitive hagfish, produce huge quantities of slime when caught. Place one in a bucket of water, and after a while the bucket will only contain

Pink ling
Genypterus tigerinus
80 cm
5 kg

Deep to coastal species around the southern half of Australia, particularly Victoria and Tasmania. Related to the true cods of the northern hemisphere. The pink ling is renowned for exuding incredible quantities of body slime when caught.

Slimy mackerel
(Blue mackerel)
Scomber australasicus
65 cm
2.2 kg

Larger fish common offshore around the southern half of Australia. Juveniles found inside estuaries, bays and inlets.

the fish plus a white, jelly-like, thick mucus. The pink ling, an excellent eating fish, also has the same sort of defensive mechanism, a rather distasteful fact that is not conveyed to seafood consumers for obvious reasons.

STREAMLINING

The other important function of body slime would appear to be that of reducing friction between the body and water, especially in fast-swimming fish. The soluble polymers within the slime may have the effect of stabilising laminar flow at high velocities (in other words, reducing turbulence at the body surface), and thereby reducing drag – the enemy of speed. This reduction of friction has been proven under laboratory conditions in live fish, and by comparing drag at the surface of freshly killed fish with the surface of wax models. I've often noticed that tuna and marlin produce liberal quantities of body mucus and no doubt streamlining is a factor there. But having noted this aspect of slime, it should be stressed that much work still needs to be done in the area of slime biology.

OTHER FUNCTIONS

A discussion of fish slime and its functions would not be complete without mentioning one fascinating use of mucus in some species of reef fishes. Each night, many species of parrot fish swim into a cave, and loosely wrap themselves in an elaborate kind of filamentous cocoon made from mucus. The function of this is not clear, although it may be to protect them from marauding predators such as moray eels. Whatever the reason, leaving a gap for water to be drawn in and out, they tuck themselves in for the night and go to sleep.

Much of the information here was derived from an excellent paper: "Functions for fish mucus", by KL Shephard, published in *Reviews in Fish Biology and Fisheries, Volume 4*, pages 401-429.

The Colour Purple

Can billfish see colour? A good question I'm sure you'll agree, and one guaranteed to attract an attentive audience, especially among the many devotees of the art of billfish fishing with lures. Some exciting and innovative research in this field has been happening in Australia, but before bringing you up to date with the progress of that work, it will pay to run through a little history of research into colour vision of pelagic fishes, including billfish.

The most definitive studies on this topic, and until recently, the only studies, were carried out aboard Japanese longline vessels in the late 1970s. Various tuna and billfish species were caught, their eyes dissected out and their optic nerves hooked up to cathode ray oscilloscopes. Monochromatic (single colour) light covering the whole visible spectrum was then shone onto the retinas and records were made of the responses. Theoretically, if the retina responded to a particular colour, it would cause an impulse in the optic nerve which would be then show on the screen. The results of tests on blue marlin, black marlin, yellowfin and bigeye tuna and albacore were very consistent, and led to the same conclusion for all these species. There was no response in the red, orange and yellow part of the spectrum, meaning that the only colours these fishes could apparently see were greens, blues and purples — end of story; or so it seemed for the next 20 years or so. Enter a research team from the

Can billfish and other pelagic species see colour? Lure manufacturers and users are convinced they can.

University of Queensland, Professor Jack Pettigrew, Dr Justin Marshall and Kerstin Fritsches. Professor Pettigrew had observed some billfish being tagged and released and what interested him was the lighting up of sailfish and marlin when excited, strongly suggesting that ultraviolet colours were involved. Did this mean that billfish, as was known for some other fishes, could see ultraviolet, and further, what other colours could they see?

With the help of gamefish anglers in New South Wales, Queensland and Hawaii, Ph.D student Kerstin Fritsches was able to obtain fresh eyes from several billfish species to investigate these questions. The first question of whether marlin can see ultraviolet light was quickly answered. When pure light was shone through lenses of marlin and sailfish, it was clearly shown that the ultraviolet wavelengths were completely blocked. Next, Kerstin found, under electron microscopes, that billfish retinas contained cells called rods, which can see black and white, plus two different types of cone cells which in other animals indicate a full range of colour vision! So how does this finding sit with the earlier work? Well, it turns out that the retina of billfish is not uniform with respect to the distribution of colour sensing cone cells. In fact, the area which is most likely to see colour is on the bottom of the retina, meaning

Many shallow water reef species are beautifully coloured in all the hues of the rainbow. Surely this indicates that they can see colours as well?

that colours can be seen when the billfish is looking up into bright light, but not when it is looking down into dim light or straight ahead. It is therefore likely that the original Japanese investigators shone light onto the middle of the retina, an area which cannot see all colours.

Dr Kerstin Fritsches of the University of Queensland is conducting ground breaking research on colour vision in fish.

Returning to UV vision, the flashing light which billfish reflect when excited was found to contain a strong ultraviolet component, but if billfish themselves cannot see this part of the spectrum, why should this be so? It could, of course, be by chance, but in Nature, natural selection usually produces effects such as this for a reason. The Queensland University team therefore looked at other fishes within the ecological gambit of billfish, and sure enough, found that one baitfish species in particular, the blue mackerel, was indeed able to see ultraviolet light. It is therefore likely that billfish are using the UV light against some of their prey species. They may be doing this either by startling the baitfish, or by distorting their own silhouette to confuse the prey. Either way, the object of the exercise is to gain that split second of advantage over the fleeing prey.

One other aspect of billfish vision being investigated by Kerstin is proving very interesting. It turns out that billfish have large eyes not so much for sharp vision, but for extreme sensitivity. In other words, their eyes are geared for seeing well while moving at high speeds and for hunting in low light conditions. This finding adds weight to some results I obtained while tracking black marlin at night. They were certainly active after dark, and under a starlit night sky, it is highly likely that they are still able to feed quite effectively.

Stripe Me Lucky

Why do many species of fish have stripes? Let's firstly look at the phenomenon of stripes on fish. Many, many species have vertical stripes or bars and one only needs to scan the index of common names in any fish book to discover this. We have striped everything, including anglerfish, boxfish, bass, catfish, goby, blenny, marlin, puffer, surgeon, sweetlips and tuna. There are heaps of barred, banded and striated species, as well as yellow-striped, black-striped, three-striped, six, seven and eight-striped fish. Continuing the theme, we even tend to name striped fish after land-based counterparts, such as the tigerfish, tiger shark, tiger blenny, zebra fish, zebra dartfish, zebra lionfish and, of course, even the good old "footballer", which is a striped version of the coral trout.

The exquisitely marked humped batfish, Platax batavianus. *A classic example of stripes being used to break up the outline of the fish.*

Some species are permanently striped, while some only show bars in certain environments or light conditions, or when excited – for example when hunting, feeding, spawning, fleeing or hiding. Fewer species have permanent horizontal bars, although some horizontally striped species such as skipjack or bonito show vertical bars when in a feeding frenzy. The juveniles of many species, especially the large trevally family, show prominent vertical bars, but these disappear in adulthood, or at least become transient or rarely apparent, such as for the golden trevally.

Stripes on fish are obviously there for a reason (or a combination of reasons). The process of natural selection has resulted in this particular body patterning on many occasions, and we can presume for starters that the reasons are visual ones. The most likely reasons being to either signal to other fish of the same or different species, or to hide from them in some way. It's not hard to visualise that prominently barred fish which live around reefs and weed use stripes to break up their outline against the background, and this would be especially true for juveniles. But what about open water or pelagic fishes? Why do so many of them show this feature, albeit fleetingly?

Well-known examples of barred predators would include striped, blue and sometimes black marlin, sailfish, wahoo and mackerels, while yellowfin tuna are marked prominently with

Zebra fish
Girella zebra
53 cm
3.5 kg

Confined to southern Australian coastal reefs, especially South Australia. Closely related to the common luderick or blackfish, the zebra fish has by far the most prominent vertical stripes of the family.

vertical striations on their bellies and flanks. Observations of dolphin fish from underwater viewing posts and of mackerel tuna in captivity have revealed that they too can produce vertical banding patterns when aroused or feeding. Sometimes the bars persist for a while after feeding and, on this point, even for those species showing more or less permanent barring, the bars always intensify when the fish are excited, usually when feeding. This may also happen when the fish is hooked.

Vertical barring does not seem to be so universal among the bait-fishes, but certainly many are at least capable of producing transient vertical bars on their bodies. For example, both slimy mackerel and yellowtail scad will become markedly barred after settling down in a covered live-bait tank. Flying fish too can often be seen with several vertical purple stripes on their flanks. But why this seemingly universal pattern? Hemingway, in The *Old Man and the Sea*, was also curious about this phenomenon when he wrote, from the viewpoint of the old fisherman, Santiago:

"I would like to fly very slowly at two hundred fathoms high and see the fish from above. In the turtle boats I was in the cross-trees of the mast-head and even at that height I saw much. The dolphin [fish] looked greener from there and you can see their stripes and their purple spots and you can see all of the school as they swim. Why is it that all the fast-moving fish of the dark current have purple backs and usually purple stripes or spots? The dolphin looks green of course because he is really golden. But when he comes to feed, truly hungry, purple stripes show in his sides as on a marlin. Can it be anger, or the greater speed he makes that brings them out?"

Not being able to really answer these questions from the fishes' point of view, quite probably the barring (and lighting up) of predators serves to break up the outline of the swooping hunter, confusing the prey just long enough for the fatal strike. Perhaps in species such as skipjack it signals other members of the school and triggers the feeding frenzy which in turn panics the baitfish, allowing easier pickings. In the case of the baitfish themselves, the vertical stripes may also serve to camouflage, especially within a large school of seething fishes: where does one fish end and another begin?

All very interesting to think about, inevitably leading to other questions, such as: should we put stripes on lures or not? Do stripes turn predators on, do they make the lure more lifelike, or do they actually make it harder to see? The more we know, the more questions are revealed.

Juvenile golden trevally, Gnathanodon speciosus, showing prominent stripes. Why is this patterning so widespread in the world of fish?

The Oldest Fish?

"Tish the fish dies at 43." This great headline, buried on page 25 of a Sydney daily newspaper, naturally caught my eye. After all, fish don't make the news very often. The item, in full, read as follows: LONDON: The world's oldest goldfish in captivity died – at the age of 43 – in its home in Thirsk, north England. Tish the goldfish began life as a "roll-a-penny" prize won by seven-year-old Peter Hand at a fairground in 1956. As Peter grew up and married, Tish, measuring 11.5 cm, aged gracefully, turning from brilliant orange to silver, and becoming something of a celebrity. "I don't think we will be getting another one," said Hilda Hand, 72, after the goldfish's death. "We couldn't replace Tish: he was part of the family."

The article raised all sorts of questions in my mind. How did they know it was the oldest goldfish in captivity? (Had anyone done a major survey in China and Japan for example, where goldfish are often revered?) Did the term "in captivity" imply that there were older goldfish known in the wild? And, hey, how did they know what age Tish was when he (or she) was won as a prize? After all, it was presumably not an egg or a larva at the time, so it might have already been 43, or 143 for that matter. And I loved the fact that Tish had died "at home". Nice to know that the family had not sent him (or her) packing to the old fish home to spend his (or her) last days gazing out at an unfamiliar world from a strange fish bowl. And, lastly, did Tish really die of

Orange roughy
Hoplostesis atlanticus
52 cm
1.5 kg

A very deep water species which occurs in higher latitudes, mostly on the slopes of sea mounts. Caught in huge quantities in the 1980s, it was soon discovered that the great schools of orange roughy were largely composed of very old fish – most aged 100 years old or more.

old age, or did someone finally put too much fish food into the bowl after all that time? If Tish was such a celebrity, surely the body would have been donated to science so that we might be able to unlock the secrets of longevity. I await the results of the autopsy with ba(i)ted breath.

Seriously though, this bittersweet, touching little story does raise the question of longevity in fish – a question that has fascinated aquarists and fisheries biologists for many years.

Believe it or not, the idea that fish might actually be immortal was considered a topic of serious scientific debate as recently as the 1930s. The reason for this rather bizarre speculation is that fish keep growing all their lives, whereas mammals stop growing when they reach sexual maturity. Cessation of growth was taken as being the start of senescence (growing old), and since fish do not have such a point in their lives, why, they must be potentially immortal. Naturally, such a possibility has enormous ramifications for human longevity, that is, of course, if it were true. (Early medical researchers were also interested in the secret of longevity in giant tortoises and how this might be applied to humans – forever seeking the elusive fountain of youth.)

The idea that fish may be immortal was shown to be erroneous by simple but detailed laboratory observations of guppies over many generations. These studies showed that even though average survival could be increased by providing better living conditions, the maximum lifespan of the species remained the same. In other words, fish do age even though growth continues past maturity. There are quite a few species of fish that live for very short periods, some for only a year. In these species, while survival up to 10 months may be as high as 100 per cent, after then mortality increases rapidly, along with signs of senescence, or old age. These include gradual curvature of the spine, clouding of the cornea and a "generally ragged appearance" (I think I know the feeling). In other words, fish also become geriatric. On the other hand, known long-lived species such as sturgeons apparently show relatively constant mortality rates of 7-10 per cent per year until they reach about 30 years old, after which the mortality rate rises rapidly each year. This increase in mortality in later life is termed "senescent mortality".

A few species do show a sharp increase in mortality after maturity or, more to the point, following spawning. The best known example of this phenomenon would be the Pacific salmon which after spending years at sea returns to its birth streams to complete the cycle. After spawning, nearly all of the males undergo a rapid deterioration, dying quite quickly. Females may survive to spawn again. The signs preceding death suggest that this process may be a form of greatly accelerated ageing, and has attracted considerable scientific attention. Certainly, migrating salmon show signs of deterioration of many internal organs.

Finally, it seems that fishes with high metabolic rates, such as tuna and possibly billfish, may not live long lives. Skipjack tuna, for example, are known to be highly strung bundles of energy, swimming constantly at high speed and spawning every day or so for months on end. Is it any wonder that they probably only live for a few years?

AQUARIUM LONGEVITY

Another interesting source of information on how long fish live comes from keeping different species in aquaria. It should, of course, be realised that such artificial conditions do not necessarily reflect what happens in nature; for example, fish might live longer in captivity since they are fed regularly and don't have to cope with predators and the like. On the other hand, they may not live as long as normal due to stress, disease and so on. Having said that, it is still of considerable interest to see just how long some species do survive in captivity.

The only paper I am aware of on the subject was written several decades ago, but it was an excellent survey of longevity of fish from various aquaria around the world at the time. The longest times in captivity of different species makes interesting reading (age in years in brackets): sturgeon (69), grey nurse shark (24), lake trout (24), mullet (23), carp (21), Australian lungfish (22), moray eels (20), toadfish (13), English pike (10). Of course, this list may well represent how long aquaria have been in operation, and it would be very interesting to conduct a similar survey today.

AGEING STUDIES

Captive fish may give some idea of how long fish live, but there isn't an aquarium that's been standing long enough to host the champions of longevity in fish – the famous deep sea orange roughy. Estimates of maximum ages of fish are derived mainly from counting rings on bony parts, including scales and earbones (otoliths). This method has shown conclusively that your average plate-sized orange roughy would be at least 100 years old with the oldest so far panning out at about 140. Surprisingly, a lot of other species that live in deep water (hundreds of fathoms) also seem to be very long lived, the suggestion being that their metabolism at those depths and temperatures is much slower than their shallow-water cousins. So even though some fish do reach a ripe old age, and while most keep growing throughout their lives, at least we now know that the piscatorial fountain of youth is as mythical as the human variety.

Telling the Boys from the Girls

A question I am asked surprisingly often is how one can tell the sex of a fish. Look at any book on birds and you will notice immediately that males and females are often so different that two illustrations are required to show the characteristics of one species. Pick up a fish book, however, and you'll find that obvious or even subtle external differences between males and females are well and truly the exception to the rule.

Some fish do show obvious differences in the external features which indicate their sex (the scientific term is "sexual dimorphism"). The best examples of this would be sharks and rays, whose sex is easily determined by the fact that males possess a pair of claspers near their ventral fins (under the body towards the tail) while females don't. The claspers are actually used to deposit sperm into the female, all sharks and rays having internal fertilisation that allows the early development of young to take place inside the mother's body. However, only a small number of the thousands of species of bony fishes shows such sexual differentiation.

Unlike many bony fishes, sharks can be sexed externally. Male sharks (right) and rays possess organs called claspers at the base of their anal fins. Females (left) lack these appendages.

Guppies, for example, have also gone down the evolutionary track of internal fertilisation, and in this freshwater family, the male's anal fin is modified into a copulatory organ, readily identifying its sex. Other well-known cases of sexual dimorphism are shown by some of the salmonids, such as rainbow trout, the males of which develop a hooked jaw when mature. In the sea, a number of reef and inshore fishes show marked differences between the sexes, in particular the colourful wrasses. These fish are known to perform complex courtship rituals; quite possibly the brighter colour of the males is used to attract and impress the females (as is the case with birds). And a final example that comes to mind is the dolphin fish or mahi mahi. Adult male dolphin fish develop a remarkably high forehead, and are even known as bulls, whereas females retain the low head profile throughout life.

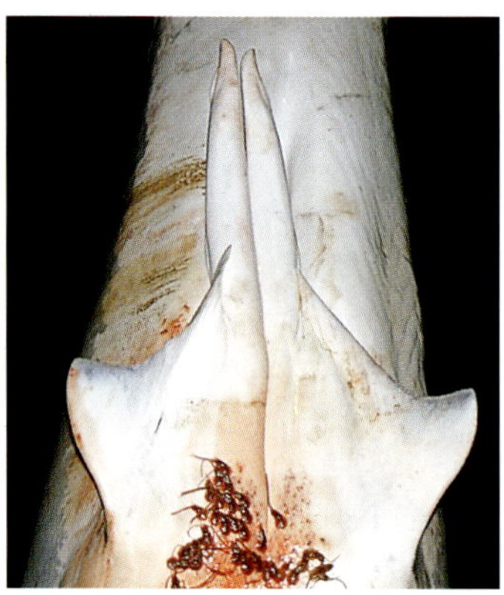

Having outlined these exceptions to the rule, it is still true that the great majority of the species of bony fishes shows no obvious external sexual differences. Which leads us back to the question: how do you tell the difference between male and female fish? Well, the only sure way in many cases is to look inside the fish. Pictures speak louder than words, so let's take a look at the photos hereabouts of dusky flathead caught during the summer spawning season on the NSW South Coast. The fish have been dissected to show the sexual organs, or gonads; ovaries for females, testes for males. Both are paired organs lying alongside the gut, and in the case of spawning fish such as these, they are quite distinctive. The female ovaries are round in cross section, and usually coloured orange or yellow. If the fish is nearing spawning condition, tiny eggs inside the ovaries will be visible, not unlike grains of fine sand. In the case pictured, the condition of the fish would be termed running-ripe, with many of the eggs having absorbed water (hydrated) and swollen perhaps 1 mm in diameter. Male fish show more angular or lumpy organs (testes) which are usually a creamy colour, and often form many segments, although in this case they are completely inflated. One telltale characteristic of the testis is that there is a distinct tube running the whole length on the outside of the organ, whereas the ovary is either solid, or when spawning has a central large space, or lumen. In the running-ripe condition, it is sometimes possible to sex live fish by squeezing gently on their bellies and seeing if either eggs or milt is extruded, but this is only possible during the relatively brief spawning periods.

Right: Ripe male dusky flathead, Platycephalus fuscus, *showing cream coloured, lobed testes. Left: Female dusky flathead in spawning (running-ripe) condition. The orange ovaries contain millions of eggs.*

While on the subject of fish reproduction, it often surprises people to hear about the incredible numbers of eggs produced by different species. For example, a dusky flathead of about 2 kg would contain about 2 million eggs. Not surprisingly, the larger the fish, the greater number of eggs, right up to the mighty black marlin. All black marlin over about 180 kg are females, and a fish around the magic 1000 pound (454 kg) mark would produce at least 100 million eggs. (By the way, if you do happen to catch running-ripe fish of both sexes, it is a

Male (top) and female (bottom) blue groper, Achoerodus viridus, *a good example of sexual dimorphism among fishes.*

good idea to mix the eggs and milt and return them to the sea. There's no guarantee that the eggs will be fertilised, but if they are, you will be doing your bit for the next generation.)

It can be an interesting and instructive exercise to sex your catch, and it is often a surprise to find a marked imbalance of sex ratios of various species of fish taken at certain times of the year. In these cases, it seems that the boys and girls prefer their own company for most of the year, only coming together for the annual spawning ritual.

Like a Fish out of Water

It probably doesn't enter the average consciousness very often, but it is sometimes sobering to remember that our ancestors, at least in the fairly distant past, were fish. Cartoonists like to depict the momentous occasion when a fish first crawled out of that ancestral pond onto dry land. My favourite version being that of the fabulous Gary Larson showing the first fish "astronauts" venturing into the unknown complete with a crude spaceship and helmets, filled with water, of course. Or how about his other classic titled "Great Moments in Evolution" showing three young fish with a baseball bat about to retrieve their ball that has been hit out of their pond. Yeah, maybe that's how it all began.

Of course, the real story of evolution from aquatic to terrestrial environments would have been far more gradual. How and, more importantly, why it happened is anybody's guess, but that hasn't stopped speculation. The most likely scenario, or at least the one favoured by most evolutionists, is that of fish gradually becoming more and more terrestrial in an environment of increasing drought. This model demands that the first land-walkers were freshwater fish, leaving drying pools over time, with natural selection favouring a predisposition to survive for longer and longer periods out of water. The whole process would have taken perhaps millions of years, with slow evolution into the first amphibians (not unlike salamanders) which still needed to return to the water to reproduce. But what about today? Are there any examples of fish that might be in the process of evolving into independent air-breathing land-walkers? The short answer is, well, maybe.

There are, in fact, quite a few species of fishes that could be termed "amphibious", at least for short durations. In a fascinating paper in *Reviews in Fish Biology and*

Snakehead
Channa sp
90cm
8 kg (some to 30 kg)

The snakeheads are a large freshwater group distributed through tropical Asia and Africa. They possess elaborate internal channels over their gill chambers which allow them to breathe air and therefore make extensive cross-country migrations.

A marine clingfish, Aspasmogasgter tasmaniensis, *often found clinging to the undersides of coastal rocks after the tide has gone out.*

Fisheries, MJ Sayer and J Davenport have summarised what is known about the subject. Amphibious fishes are those which apparently leave the water voluntarily or, as defined in the paper, fishes which "spend periods of time out of the water, on or above the ground surface, as normal parts of their life histories". This definition includes not only those fishes that crawl or jump out of the water, but also those which are habitually stranded on outgoing tides, surviving to return to the water on the next tide.

Probably the most widely recognised amphibious fishes are the endearing mudskippers of the tropics. Anyone who has fished among the mangrove creeks and mudflats of northern Australia will have seen these fascinating little bug-eyed fishes slithering and hopping about their muddy homes, eyes raised comically high on their heads like cartoon aliens. Mudskippers seem to have developed this peculiar life habit in order to exploit terrestrial food not available to competing fishes, and their skittering around chasing insects on the mudflats certainly supports this theory.

Other familiar amphibious fishes include members of the snakehead family of the Indian subcontinent, as well as most of the freshwater eels. Any farmer who has dug a dam will attest to the inevitability of eels eventually finding their way there. This is achieved by slithering over wet ground on dewy or humid nights, often over impressive distances. The reasons for this behaviour are not fully understood, but it is obviously an effective way of dispersal, especially when the home pool is evaporating and becoming more than a little crowded.

A range of motivations for fish leaving water, at least temporarily, have been put forward. These include positive reasons such as the exploitation of food, as in the case of mudskippers, but mostly include negative responses such as escape from one thing or another, whether it be physico/chemical or biological. Physico/chemical reasons mostly relate to drying and shrinking pools, and would include decreasing oxygen levels, increasing concentrations of toxic chemical byproducts and/or increasing temperatures or pH (acidity). Biological factors may be equally compelling reasons to leave the watery world, in particular to escape from either competition from other fish or from predators. It certainly does make sense to be able to move into a completely alien environment to avoid being eaten, a tactic favoured by the intertidal and secretive clingfishes, many of which cling to the underside of damp rocks above the high tide zone. Some species of fish have adopted this tactic not for themselves, but for their offspring. The male and female of one species of freshwater fish in Guyana (a characid) actually leap onto a leaf above the water to spawn and deposit their eggs away from lurking predators.

Some coastal saltwater fishes have evolved a similar tactic of reproduction, at least until their eggs hatch and the young can fend for themselves. Although we don't appear to have any examples of these in Australia, the north American grunion is certainly a famous exponent of this feat. On certain predictable times of the year, grunions mass along beaches of southern California, flinging themselves ahead of the flooding tide to deposit their eggs and milt. This fishy ritual has led to a ritual human exploitation of grunion, but the fertilised eggs left behind, safe from predation, always ensure another generation will follow.

So are any of these examples likely to evolve into terrestrial air-breathers. Possibly, but if I were to put money on the most likely candidates, I would pick the lungfishes. These true air-breathing fishes, species of which live on each of the southern continents of Australia, Africa and South America, are closely related to our likely lobe-finned ancestors. Given the right conditions and plenty of time (and a game of underwater baseball), it's not too hard to imagine these ancient relatives of ours beginning the whole process all over again.

Lungfish are capable of breathing air, and are thought to resemble the ancestors of terrestrial vertebrates. Pictured is the Australian lungfish, Neoceratodus forsteri.

Raking It In

The anatomy of fish is a fascinating source of study in its own right. While fish are vertebrates, and like all vertebrates, possess familiar structures and organs such as a spinal cord, gut, liver and kidneys, they also have certain features that are completely alien to air-breathers like us. The possession of gills is obviously a case in point. To extract oxygen from its dissolved state, gills consist of vast numbers of filaments, each supplied with blood vessels, maximising surface area and allowing rapid gaseous exchange. But I want to explain gill rakers.

Most fish species have varying numbers of hard and sometimes sharp projections on the inner edges of the gill arches, and in many cases they do resemble the prongs of a garden rake; hence, gill rakers. These projections, when viewed from the front of a fish with its mouth open, form a kind of overlapping cage or mesh. The main function of these gill rakers is to trap or filter food that is then taken into the mouth. A fish cannot help taking in water with food, so the trick is to swallow the food while at the same time expelling most of the water from the gill openings. Gill rakers do this job very well by sieving out solids, and in fact the spacings between the rakers usually bear a close relationship with the size of the food items normally eaten.

As it happens, gill rakers are a useful feature in identifying many species of fish. For example, when I caught a large tuna off Bermagui some years back, its identification was a bit of a problem. It wasn't a yellowfin since it did not possess long sickles, but it could have been either a southern bluefin or a bigeye tuna. Consulting some up-to-date references, I found that bigeye tuna have 23 to 31 gill rakers while southern bluefin have 31 to 40. My fish had 25, so bigeye it was.

Very small food items found in the stomach of a longtail tuna, Thunnus tonggol. *Note that fish, crustaceans and squid are all about the same size, presumably seived from the water by the tuna's gill rakers.*

Because plankton is abundant, many species of fish have evolved methods for extracting various components of the plankton from water. Filter-feeding fish all have long, densely packed gill rakers, some resembling fine gauze. Herring may have up to 25 rakers per centimetre to strain copepods (small crustaceans) from the water; an obviously efficient method since up to 60,000 copepods have been found in the stomach of a single herring.

Predatory fish tend to have far fewer, but much stouter gill rakers, and these are the ones which can cause some problems with injuries to humans when putting one's hands into the gill cavity. Teeth are often blamed for cutting line, but the rasping of gill rakers can easily achieve the same result, hence the preferred use of wire trace for many of the larger predatory fishes.

Interestingly, one of the conspicuous fish groups that does not possess any gill rakers at all are the billfishes (marlin, sailfish and spearfish). Being apex predators, they tend to dine on large prey, so presumably have dispensed with the need to sieve their food. However, one very strange finding from studying gut contents of blue marlin in Hawaii was that they often contained very small food items, such as crab larvae, tiny triggerfish and stomatopods, or prawn killers. One explanation of this finding is that large blue marlin are capable of picking small, individual food items from the water rather than swimming through clouds of prey and filtering them out. The idea is that when larger prey are scarce, marlin may obtain at least some of their energy requirements by lazily picking away at passing morsels. To me, this does seem weird, though, and maybe there is another explanation waiting to be discovered.

A cleaner shrimp gets to work on the gill arches and rakers of a large cod. Cleaner shrimps and fish perform an important role in ridding these areas of parasites.

I can't finish on gill rakers without relaying the following true story. A biologist friend of mine, who will remain unnamed but who is based in Cairns, has the long-standing, unenviable reputation of being incredibly accident-prone. You name it, it's happened to him, but my favourite story about him involved, believe it or not, gill rakers. One evening while cleaning mackerel at the back of a boat on the Great Barrier Reef, all was well with the world. Our intrepid biologist was dousing some fillets in the briny while perched on the marlin board, at the same time, turning his head to chat with others onboard. His attention returned quickly to his own anatomy when a huge potato cod that had been attracted to the boat not only engulfed the

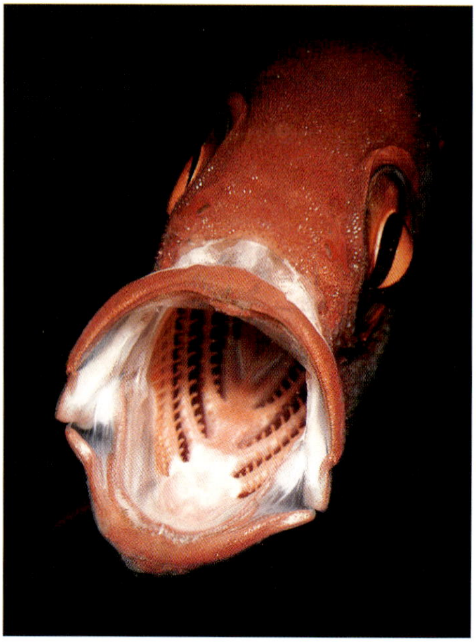

Open wide. The comb-like gill rakers of this glasseye, Priacanthus blochii, *can be easily seen along each gill arch. Rakers are used by many species of fish to strain small food items.*

fillets, but also his arms, right up to the elbows. This sudden intrusion on his personal space caused him to resume an upright stance rather quickly, and as he did so, the cod, having been lifted half clear of the water, slowly slipped back, deeply scouring our hero's arm in the process with its sharp, finger-like gill rakers.

So what should be done to avoid being injured by gill rakers of large fish? The best advice is to wear gloves if hoisting one by the gills, and, oh yes, when washing fish and hands at the back of a boat, please pay attention to what might be swimming up the berley trail.

The Amazing Argonauts

Game fishing captain Lloyd Williams of Esprit Charters in Bermagui relates a fascinating story of one of those once-in-a-lifetime chance observations. He was fishing as usual off the far South Coast of New South Wales on a rare clear, glassy day. He hadn't noticed any bird life to speak of, so was attracted to a small group of albatross and mutton birds (shearwaters) on the water. On approaching, he spotted what seemed to be a white trap-buoy about a metre in diameter floating in the midst of the birds, but closer inspection with a prod of a gaff showed that the ball was not solid, but rather a tight aggregation of animals. Ink-squirting paper nautiluses, to be exact. He managed to extract one of the beautiful creatures for inspection, having only ever seen the delicate, empty shells before. Inside was a small, short-tentacled octopus, so he contacted me to try to find out more about this rare observation.

The paper nautilus is indeed a species of octopus that is especially known to come from the far South Coast of New South Wales. Only females of the species form the thin, parchment-like shell, half of which is secreted by each of two large, flattened tentacles, then brought together and fused. The shell is only formed to protect the eggs for a time, after which it is discarded. Males of the paper nautilus are very small, free-swimming animals. Like other octopuses, one of their tentacles is modified as a

The exquisite emperor or tiger nautilus, Nautilus pompilius, *the only living squid-like animal to retain a permanent external shell.*

copulatory organ, but in this case the organ breaks off inside the female, after which the male dies (in some sort of post-coital bliss, I would like to think). The female then fertilises her eggs at leisure from the sperm stored in the organ, often a given female containing as many as eight such sacrificed "arms". (As an interesting aside, early this century, scientists observing these male appendages inside females mistakenly took them to be some form of parasite, even giving them their own Latin name.)

Lloyd had been amazed at how the individual animals had arranged themselves so uniformly to form an apparent sphere, and thought at the time that this might be a form of protection from the birds. This could well be the case, although I have read accounts of marauding yellowtail kingfish smashing into rafts of these little shelled animals, and of snapper gorging themselves on the soft bodies, mainly after storms. So the thin shells would not appear to be much of a defence. Another equally plausible explanation for the balling behaviour could be a mass spawning event, with perhaps the naked males being hidden inside the perimeter of the shells. In any case, this was a fascinating phenomenon, which Lloyd was indeed fortunate to witness.

The extraordinary paper nautilus, Argonauta argo, *an octopus which secretes a protective egg case.*

This case of an octopus with a shell leads us nicely into a brief look at other members of the group that bear shells. To begin with, squids, octopuses and their kin are all molluscs – cephalopods, to be precise. Most have long ago lost their shells, although squids, and especially cuttlefish, retain an internal shell as support for their soft bodies. One very small member of the clan, Spirula, has an internal, spirally coiled shell, often seen washed up in large numbers on beaches. As noted, the paper nautilus shell is merely a temporary egg case, but there is one group of cephalopods, the nautiloids, which have retained a permanent outer shell. The true nautiluses have beautifully patterned spiral shells which contain numerous internal sequential chambers left behind as the animal grows (unlike the paper nautilus, the true nautilus only occupies the very front chamber of the shell). These chambers are filled with gas and through valves provide controlled flotation by which the nautilus can ascend and descend at will. The nautiluses are also unique in that they have no suckers on their tentacles, no ink sacs and only a simple eye, all indicators of apparent primitive traits.

Face to face with one of the most beautiful and delicate creatures in the sea, the paper nautilus.

Today, the true nautilus is the only cephalopod possessing a hard, permanent external shell. However, hundreds of millions of years ago, spiral-shelled cousins of the squids and octopuses ruled the seas. These were the ammonites, and though not very closely related to the nautilus, their long spiral shells were similarly partitioned for presumed buoyancy. Among the commonest of all fossils, the ammonites were a diverse group of animals ranging in size from 3 cm to a huge 2 metres in diameter. The animal inside such a gigantic shell, possessing a large, sharp beak and presumably resembling a giant squid (although no fossils of the soft parts remain), would undoubtedly have been a formidable beast in the Devonian seas. At the same time, another group of shelled invertebrates, the slater-like trilobites, roamed the sea floor, and it is likely that the larger ammonites were their main predators. In fact, it is thought that trilobites quickly evolved upward-looking eyes at the time of the explosion of the ammonites in response to the constant threat from above of the vertically mobile beak-bearing ammonites.

Returning to the paper nautilus, its given Latin name is *Argonauta argo*, a delightful moniker for this charming little creature. The argonauts were, of course, the sailors who crewed Jason's fabled ship, *The Argo*, in his quest for the golden fleece. The superb sculpted white shell of the paper nautilus must have evoked images of the mythical ship in the mind of the naming biologist. And who said scientists have no soul?

Sticking to The Job

Variety is the spice of life, especially if you're talking about life in the ocean. I've said it many times during my years as a biologist, but I never cease to be awe-struck by the incredible diversity of fishes in the world, and by their amazing adaptations evolved through natural selection. One such example of evolution producing truly specialised adaptations is the remora.

The remora is one of those worldwide fishes that has been known since ancient times. The Latin name "remora" actually means "to hold back", and the Greeks and Romans certainly believed that remoras could impede the smooth progress of sailing ships, or even stop them altogether. The Roman author Pliny wrote "why should our fleets and armadas at sea make such turrets on the walls and forecastles, when one little fish is able to arrest and stay our goodly and tall ships?" This feat was depicted on vases and other pottery, and even the defeat of Antony and Cleopatra by Augustus Caesar in a naval battle was attributed to a remora holding up their galley.

On his second voyage to the West Indies in 1494, Christopher Columbus observed native fishermen who used remoras with a line tied to their tail to catch turtles. When a turtle was spotted, the fishermen would release the remora to swim towards and latch onto the turtle. The line could then be carefully retrieved to reel in the catch. Australian Aborigines are also recorded as using this method. (Maybe it could even be used by an enterprising game fisherman to catch a marlin.)

Slender remora
(Sucker fish)
Echeneis naucrates
2.5 kg

Tropical to temperate distribution, around Australia to Albany in Western Australia and Bermagui in New South Wales. Mostly found clinging to larger fish, such as sharks and rays, but also can be seen free swimming near reefs. The amazing sucker on top of its head is actually a greatly modified dorsal fin.

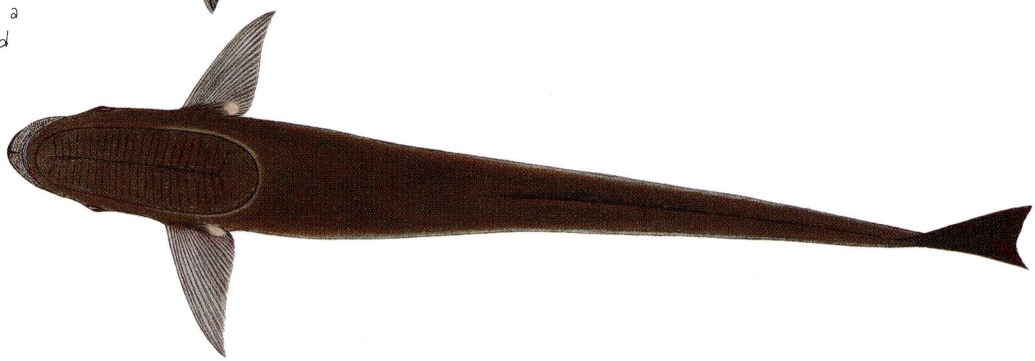

The remora attaches itself to any flat surface by means of the elaborate sucker on top of its head. Close inspection of this odd structure shows, amazingly, that it is really a modified dorsal fin, evolved into an incredibly efficient suction device. The sucker has a series of transverse plates that are free at the rear, and which can be raised and lowered (being modified rays of a dorsal fin). When the remora wants to attach itself to any object, it makes firm contact and raises the plates slightly, creating a series of vacuum chambers. The resulting adhesion is so strong that it is almost impossible to prise the fish off – the only method appears to be to slide it forward until the suction is broken. I found reference to one apparent study many years ago at the New York Aquarium where remoras were placed into buckets of water, allowed to attach on the inside, and then lifted by the tail with spring balances. The champion remora of the day had the power to lift a weight of nearly 20 kg. (I wasn't able to find out how they actually attached the lifting line to the remora – not the sort of experiment one would be permitted to conduct these days, methinks.)

The original hitch-hiker: the remora, or sucker fish, once thought to have the power to stop ships. This non-pigmented species lives inside the gill cover of large fish such as marlin.

The lovely little remora pictured here came from the inside of the gill plate of a black marlin brought to the weigh station during a major game fishing tournament at Port Stephens, New South Wales. This was because I had put out a call for any boat landing a marlin to remove any live parasites from the skin and keep them alive in a bucket. At the time, I had meant the so-called sea lice (copepod crustaceans) which are often found skittering around the surface of marlin, so I was rather surprised when several buckets were handed to me at the end of the day containing some very active remoras. In fact, remoras are not parasites at all, but rather hitch a ride for several reasons, including to gain protection from predators, to feed on morsels left over by their host, and possibly to be transported to new areas. In the case of the marlin remora, its lack of pigmentation strongly suggests that it spends nearly all its time inside the gill chamber, obviously protected and presumably getting lots of free dinner scraps.

Even though they are the ultimate hitch-hikers, remoras can actually swim quite actively themselves. Often, free-swimming remoras can be seen chasing food scraps at

A small black marlin showing two remoras clinging to its underside.

the back of boats anchored behind reefs in northern Queensland. They will take a bait readily, and put up quite a battle, and although I've always let them go, Ern Grant, in his bible, *Guide to Fishes*, reckons that skinned remora fillets are quite good eating.

A fish that bears a striking resemblance to the remora, and which is also good to eat, is the cobia, itself a worldwide species. Neither the remora nor the cobia has any obvious, close relatives, but many scientists suspect that the remora and the cobia may well have evolved from a common ancestor. Cobia have a habit of swimming very close to sharks and manta rays, so it is not a giant leap of speculation to imagine the link becoming much closer over evolutionary time scales.

There are about six species of remoras worldwide, some being quite specialised, such as the whale remora, which only attaches to dolphins and whales, and the shark remora, which prefers the company of large sharks and rays. The spearfish remora, as the name suggests, lives on the surface of marlins, while the white remora dwells inside the gill cavity of marlin, and also of manta rays and some sharks. The largest species, simply called the "remora", grows to more than 60 cm, and can often be seen attached in pairs to the lower surface of manta rays. It seems they balance their host, since the two are always equally spaced, looking for all the world like big, black fuel tanks under a stealth bomber.

The Empty Nest

Do some fish care for their young? At the outset, it should be said that the great majority of fishes do not appear to be at all concerned over the fate of their offspring, especially most species producing hundreds of thousands or even millions of eggs at a time. The whole point of broadcast spawning, as it is called, is to cast as many eggs as possible to the vagaries of nature, for dispersal and survival of the fittest (after all, it only takes survival of two from the millions to replace the parents). Mind you, most of the broadcast spawners at least try to ensure that their larval offspring will hatch in a friendly environment that provides shelter, food or both.

A *"pregnant" male seahorse,* Hippocampus breviceps, *giving birth to its young via muscular spasms of its pouch.*

Even though broadcast spawning and larval dispersal is the rule in the fish world, parental caring in some form or other is known in quite a number of fishes, most of which, for some reason, happen to be freshwater species. Given the enormous diversity of fishes on the planet, it is perhaps not surprising that parental care has evolved in many different ways. It may range from making sure that eggs are protected until hatching, through to guarding the young for some time after hatching. The full spectrum of parental care has been classified into a wide array of patterns but let's look briefly at the main reproductive strategies, from no care at all, to the best known examples of true parental care (with examples which come to mind in brackets).

Non-guarders

This category includes all species that broadcast spawn in open water (tuna, yellowtail kingfish, mackerel, trevally, cod, drums and so on), on or near the bottom (trout, salmon) or attach their eggs to plants, then ignore them (garfish, ballyhoo).

Brood hiders

These are species that conceal their eggs for protection, but do not tend or mind them (a good example is the North American grunion, which buries its eggs in sand at a peak high tide, where they are safe from predators until hatching on the next month's peak high tide).

A pair of nest-guarding clownfish, Amphiprion polymnus, *taking good care of their clutch of eggs, laid on an old drink can!*

GUARDERS

Guarding is where parental care really comes into play. Guarders are species that protect and defend their eggs until hatching. Eggs may be deposited, or stuck to, bottom gravel or rocks, on plants, in caves or holes, or, especially, in carefully constructed nests.

NEST SPAWNERS

These are specialised guarding species that construct nests as simple as a scooped-out bed of gravel or as elaborate as tunnels and half-pipes built from glued-together scraps of plant material (the most famous nesting fish are the British sticklebacks).

BEARERS

While all of the preceding types of care involve egg protection, bearers are species which not only hold their eggs on or in their bodies prior to hatching, but then, in many cases, protect the young fish after hatching. The seahorse is a famous example of this strategy, in which the female lays her eggs into a pouch on the male where they develop until hatching. At that time, the male actually "gives birth" by ejecting the babies in a series of muscular spasms. Quite a number of species hold their eggs in their mouths until they hatch. This technique is called mouth brooding, and has evolved in several unrelated Australian freshwater fishes including fork-tailed catfish and the saratoga. In the latter case, the baby fish stay very close to the mother after hatching, darting inside her mouth again whenever danger threatens. (It has always seemed a little ironic to me that this most advanced form of parental care in fish has developed in the saratoga, which is regarded as one of the more primitive fishes.)

When we think of parental care in animals, we usually think of the parents not only protecting their young, but providing them with food as well. There are no known examples of parent fish actually gathering food for their offspring, however there is one rather bizarre case which does qualify as feeding. The young of the discus fish (a freshwater cichlid) stay close to their parents, which remain as a pair for some time after their eggs hatch. The baby fish spend much of their early life feeding off the body slime of both the male and female parents, and from experiments it has been shown that the young actually prefer the slime to other food items. One interesting observation is that the parents seem to take turns in feeding their progeny, but apparently periodically tiring from the constant picking, send the babies off to their partner with a shake of the body.

As you would appreciate, broadcast spawners such as billfishes, tunas and the like are very unlikely candidates for parental carers since their eggs would not hatch until they had been carried far away on the oceanic currents. A friend once reported to me an observation of what appeared to be a sailfish minding a group of baby sailfish. The probable explanation in this case, though, is that the baby "sailfish" were much more likely to have been a school of small longtom (needlefish) or garfish, and that the "parent" was circling around with dinner, rather than parental instincts, on its mind.

A fabulous shot of a marine mouth-brooding cardinal fish, Cheilodipterus macrodon. *Note the silvery eggs protected by the ferocious looking teeth.*

School's In

A seething, swarming school of fish can be a mesmerising sight. Sometimes so tightly packed together, the mass of fish can appear to move as a single, living entity with its own flowing purpose and direction.

Many fish species school. In fact, it is estimated that as many as 80 per cent of all species form schools during at least part of their life cycles, and about 25 per cent spend their whole lives surrounded by friends and relatives. There are many different kinds of schools. Some are very regular, containing individuals of the same size, while some may contain other species, although those are usually similar in size and shape to species forming the main school. Schools may consist of very few individuals (often called pods) but at the other end of the scale, single schools of fish like herring and anchovies have been estimated to contain several billion individuals. Schools may disperse and re-form on a regular basis, with many species only schooling during the day and scattering at night; and when they reassemble, the same fish don't necessarily form the same schools.

Schooling behaviour can be a decided advantage in avoiding predators, unless, that is, the predator has two legs.

On the other hand, in some species members of a school may be very closely related. For example, it has been found that yellowfin tuna schools often contain

siblings (ie, fish hatched from the same batch of eggs), which means that they have remained together for at least their first few years of life.

A tightly bunched school of striped catfish, Plotosus lineatus. *In species such as this, the schooling instinct is extremely strong.*

Technically, gatherings, or aggregations of fish, can be classified into two categories: shoals and schools. A shoal contains fish that congregate but which are only loosely coordinated, whereas a school is much more structured, containing individuals that are quite evenly spaced, and which, to a large extent, swim parallel and in the same direction in a synchronised fashion. It is always difficult to generalise, but feeding or spawning aggregations of fish can usually be classified as shoals, while synchronised schooling is usually associated with predator avoidance.

Why is schooling or shoaling such a common phenomenon in fish? Obviously it must confer advantages for survival, and the main advantage of organised or synchronised schooling appears to be predator avoidance.

Avoiding predators by being part of a school is an interesting concept. After all, a school of fish would seem at first glance to be quite vulnerable to attack, but that's where the safety in numbers theory comes into play. Firstly, it is easy to imagine an attacking predatory fish or marine mammal becoming suddenly disoriented or confused when hundreds of bait-fish shower in all directions. A predator needs to focus on one individual (like a cheetah cutting a gazelle from the herd), but in the case of schooling fish, this becomes difficult – so many choices, so little time to decide. Secondly, most bait-fishes are silver, so that when they scatter in all directions, a shower of flashes adds to the predator's confusion. (Interestingly, bright flashes actually elicit a feeding response in predatory fishes, so a single, flashy lure is actually an effective bait.) And thirdly, because fish in a school respond extremely quickly to their neighbours' reactions and movements (detected by both eyesight and lateral lines), the fish on the side of the school away from the attacking predator react and flee in plenty of time to avoid the danger. Of course, this doesn't mean that the predator always goes hungry, but it does mean that there is a higher chance of escape for the majority of the school.

Mind you, the schooling habit isn't always an advantage. Large schools may be more visible from a distance. Bait-fish in the middle of a school may actually become confused themselves, becoming more rather than less prone to attack, and it is even thought that the trailing fish in a large school have to deal with depleted oxygen levels in the water, again leaving them more prone to attack. Finally, when a school of fish is on the menu of giant filter-feeders such as baleen whales and whale sharks, the tendency to bunch together would seem to be a decided disadvantage. This is because the whales and whale sharks operate by engulfing a whole school in one mighty gulp, often swimming up from below to force the bait-fish to the surface. And then there is the ultimate predator, *Homo sapiens*, who also uses schooling behaviour to great advantage. Many commercial fishing methods rely on the fact that fish school: in particular beach seining and the highly efficient method of purse seining. In these methods, the school is located visually, often from the air, and the net is deployed to encircle the whole school. Even trawling, which at first sight is a fairly random method of towing a net along the seabed, can be used to target schools or aggregations. The gemfish is a predatory fish that hunts at night but forms dense schools near the seabed during the day, making it prone to trawling. Their vulnerability is further increased by their strong tendency to aggregate at quite specific depths, making them easy targets for directed trawling. Even more specialised is the trawling method used to take orange roughy. This species forms huge spawning aggregations (shoals) on the sides of sea mounts. Commercial trawl nets are deployed hours before sinking to reach their targets, hundreds of fathoms below, neatly skimming the edges of the mountains, capturing tonnes of fish with each sweep. You may have noted a common thread here. Schooling fish by their nature are also particularly prone to overfishing, gemfish and orange roughy being prime examples. The added problem is that as the population of fish is depleted, schools still form and catch rates may still be high; until, that is, it is too late, and the fishery suddenly collapses. A school of thought to leave you with.

Schooling confers many advantages to most individuals, although there are some downsides too. Whitetip soldierfish, Myripristis vittata, *enjoy the comfort of company.*

Above: Two schools at once. A school of skipjack tuna attacks a school of smaller baitfish from below, using the surface as a "wall".

Left: Many small, schooling species are silver, creating a dazzling array of flashes that can confuse an attacking predator.

Fronts and Slicks

I came across some fascinating literature recently involving a topic that should be of interest to many saltwater anglers. An article in the journal *Trends in Evolution and Ecology* entitled "The Lost Year of the Sea Turtle" by Wiliam M Hamner caught my eye. It related to a long-standing mystery of what happens to newly hatched loggerhead turtles once they enter the sea and swim into the open ocean to become part of the pelagic fauna.

Assuming they go somewhere, the secondary, more critical problem was, what do they eat, or more to the point, how do they find food? The same questions also apply to many species of fish, as well as a wide range of ocean-going animals, including sea snakes and birds such as terns and petrels. This problem had troubled an eminent land-based herpetologist called Archie Carr, especially since turtles don't eat plankton, and they obviously can't photosynthesise. His mystery was solved during a routine flight, and his own words describe his realisation.

"Then one day I looked out of a little airplane and saw the whole surface of the sea banded with narrow weed-lines, spaced across the water with unaccountable regularity. When I got home and talked with an oceanographic colleague, I learned that the multiple driftlines were made by advection between the opposing horizontal vortices generated by steady wind, that they are called Langmuir bands, and that the phenomenon is common. For me, this eased the way toward understanding how epipelagic life [right at the surface of the ocean] can be lived by an organism unable to photosynthesise, or to sweep plankton, or to cruise away to the Gulf of Guinea when its rip disintegrates."

It is thought that baby turtles rely on current lines and slicks to find food in the open ocean.

Carr too, of course, "discovered" current lines, or slicks as they are also called, and in that instant, realised that his turtles could survive by finding these food-laden slicks, and moving from one to another.

It may surprise some to learn that the surface layer of the open ocean, distant from nutrient-producing land masses, is relatively barren; in fact a sort of marine desert. Yet we know that fish and various air-breathing animals occur out there, routinely moving through these vast wastelands. Planktonic animals and plants, the basic building blocks of marine food chains, also tend to be concentrated near land masses, but are also dispersed across oceans, albeit, thinly. It is along fronts, current lines or slicks that planktonic life concentrates, thereby forming the basis for whole biological communities, and the means of survival for travelling or drifting nomadic animals. Often, fronts themselves are areas where unlike water masses meet, perhaps of different temperatures, salinity or density. Down-welling, driven by the wind, often occurs along these boundaries, and the whole line does in fact become calmer than surrounding waters, hence the apt description of a "slick".

Structure in the otherwise featureless ocean. Logs and debris accumulate along fronts, providing shelter and food for whole communities.

The resultant rich assemblages are transient, and will break up in high winds or seas, but they form often enough and last long enough for many organisms to have evolved the capacity to take advantage of the phenomenon. Many sea birds rely on fronts to find small planktivorous fishes, while sea snakes, which generally drift rather than swim, use them to aggregate for feeding and for mating. An early observation by WP Lowe in 1932 referred to a concentrated line of a species of sea snake in the Strait of Malacca that was 3 metres wide and an amazing 96 km long. Later researchers surmised that this was a concentration caused by horizontally convergent water masses forming a classic front or slick.

Pelagic fishes, which may spawn in the open ocean, usually produce huge numbers of eggs. A large female marlin may produce over 200 million eggs, while skipjack (striped) tuna spawn several hundred thousand eggs every day or so for months at a time. It is obvious that there must be massive mortality of either eggs or early larvae of such fish, otherwise they would fill the ocean. Many scientists believe that there is massive mortality of larvae, and that this must occur due to starvation in the marine deserts between fronts. Interestingly, this theory has never really been proven. A plankton survey off the Hawaiian island of Lanai in 1956 did find some dead larvae of the frigate mackerel (*Auxis thazard*), but there is no

Convergence of unlike water masses creates slicks where debris accumulates. A good place to wet a line or troll a lure.

confirmed report in the literature of a single case of mass mortality of larvae of any pelagic species. This is not to say that it doesn't occur; it probably does, but the chances of a survey vessel being on the spot just before dead larvae sink or decompose would not be high.

Once plankton, fish larvae, seaweeds, jellyfish and the myriad other assorted drifting organisms and objects have aggregated to form a surface oriented community along a slick, larger and more mobile predators will, of course, take advantage of this veritable smorgasbord. Tuna such as albacore and yellowfin tuna are known to concentrate along fronts, and satellite technology is now used to locate the telltale plankton bands and to direct fleets of long-liners (or in the past, drift-netters) to these aggregations. Most game fish anglers are also aware of this phenomenon, and trolling the edge of a current line or slick is often rewarded with a crashing strike from a blue or striped marlin.

To Sleep, Perchance to Dream

Do fish actually sleep? Fish seem somewhat alien to many of us, and two observations that tend to make people think that fish may not sleep are that they don't have eyelids, and that they often seem to swim constantly. How could one sleep under those circumstances?

Well, the short answer is that, yes, fish apparently do sleep, although some more than others, it seems. How do we know this? Simply observing fish at night, either by diving or in aquaria, shows that many species become very quiet at night, often altering their behaviour completely. On my first night dive, I was amazed to see a number of different species of fish apparently hovering in some sort of trance. It was even possible to touch them, or even carefully handle them before they suddenly snapped out of it and darted off. Some small schooling fishes, such as mullet, alter their daytime gregarious habits when the lights go out, and head separately for the bottom where they lie facing in all directions. If they are disturbed, they quickly return to the surface and reform the school. One group of fishes is quite famous for their bedtime behaviour. Every night, some species of parrot fishes (family Labridae) put themselves to bed by secreting and enveloping themselves in a loose mucous cocoon that has an opening to allow the fish to breathe. The process of secretion, and shedding the mucus, takes at least half an hour both in the evening and when they "wake up" in the morning. The purpose of

A parrot fish, Scarus microrhinos, *curls up for the fish equivalent of shuteye. Some parrot fishes "go to bed" at night by secreting a protective cocoon of body slime.*

the cocoon is unknown, but is presumably to protect them against predators, and may also act to keep telltale odours confined.

While it is true that some fishes, notably the tunas and billfishes, reputedly must keep swimming in order to pass water over their gills (in other words, they don't suck water in by opening and closing their gill covers), they too nevertheless exhibit behaviour that might be termed sleeping. Some of my own observations support this idea. During experiments tracking black marlin off the Great Barrier Reef using tags that emitted sonic pulses, it was possible to monitor the activity of these large fish around the clock. During the day, most of the marlin tended to swim at a leisurely pace, often diving and coming to the surface as well as frequently changing direction. After dusk we came to expect increased activity, both vertically and horizontally, but only for a few hours. Later – after about midnight – the marlin would become much less active, often coming right to the surface and apparently drifting with the prevailing current for periods as long as an hour or so. It is hard to interpret this behaviour as anything but sleeping, and while this observation is somewhat at odds with the continual swimming theory, it is quite possible that oxygen demand is low during these periods of inactivity, allowing the fish to grab its version of 40 winks.

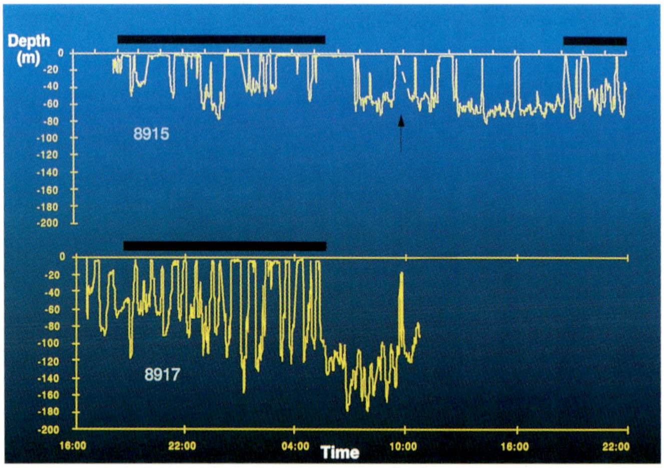

Depths of black marlin tracked with sonic tags. The black bars represent night time. Marlin tend to spend a lot of time at the surface at night, often becoming motionless for an hour or more.

Not all species sleep at night, however. Like many terrestrial animals, some fishes have developed nocturnal habits to avoid predators. Many reef fishes, for example, sleep upside down in caves during the day and become active at night, while on the other side of the coin, some predatory fishes hunt at night, perhaps taking advantage of sleepy prey. Gemfish are a good example of the latter, being a schooling, active predatory species at night, feeding on squid and other fishes in the mid-water, but remaining near the bottom during the day, presumably resting. This daytime habit has a disadvantage, however. It allowed the gemfish to be easily targeted by trawlers, leading to overfishing.

Having established that fish do indeed sleep, the next logical question would have to be, do they dream? This is not as silly a question as it might first appear.

A television documentary series on the ABC looked at the whole issue of sleep in humans and other animals such as dogs. This fascinating series noted that medical science is only just beginning to understand what sleep actually is, and why we need it so much. The conclusion reached was that we (and presumably other familiar animals such as dogs) sleep in order to dream, but we only dream for relatively short periods. There are two distinct phases of sleep, the first when brain activity declines considerably, and the second when dreaming occurs, when the brain becomes very active – as active as an awake brain, in fact. And it appears that dreaming is essential to mental wellbeing. Sleep-deprived people become quite psychotic, and there is even one genetic disorder that causes the unfortunate sufferer to gradually lose the ability to sleep, resulting in slow onset of psychosis, followed by death.

But back to fish. If, like humans, dogs, cats and horses dream, then why not fish? As far as I know, no experiments have yet been conducted to measure brainwaves of sleeping fish, so we don't yet have a definitive answer to this intriguing question. However, if I was to have a punt on what might be, and I'm entering shaky ground here, I would hazard a guess that fish may not dream, simply because their brains almost entirely lack a cerebral cortex, which is the part of the brain associated with experience and with dreaming. Then again, it's fun to imagine fish dreaming of a succulent worm, a close encounter with a shark, or last spawning season. Come to think of it, if fish do dream, do they dream in colour? And on that note, sweet dreams to all.

Up or down? Soldier-fishes, like this white tipped species, Myripristis villata, *often orient them-selves to the ceilings of caves or ledges, especially at night. In this case, it's hard to determine who is awake.*

Shark Stories

Humans' morbid fascination with sharks is probably understandable, and may even be a primitive instinct. However, it has become increasingly apparent over the past couple of decades that much of the hype surrounding sharks is a long way from reality. The dangers of shark attack and the sizes to which sharks grow are just two of the topics that have generated considerable debate through time. It also comes as a surprise to some to learn that sharks come in all shapes and sizes, and that most are, well, perfectly harmless.

Sharks come in all shapes and sizes

Shark Attack

The words strike fear and loathing into the heart of those among us who secretly hold the fear of one day ending up as the main course at a Great White's dinner party. But what are the realities of the dangers of shark attack in Australia? How many people actually do fall victim, what are the chances of survival once attacked, and which species are to blame? Fortunately, we now have some of the answers to those questions, thanks to a comprehensive computer database known as the Australian Shark Attack File (ASAF) held at Taronga Zoo in Sydney, and compiled by the Aquarium's chief technician, John West.

John began gathering the data for the ASAF in 1985, and initially drew on existing but scattered information from earlier workers including Victor Coppleson and Gilbert Whitley. He has since tracked down virtually every known incident involving a shark biting a human in Australia, and keeps all of the data on a computerised file at Taronga Zoo. It makes for fascinating reading, and some of the findings from a summary of John's are presented here.

First of all the criteria for inclusion in the ASAF are quite clear, viz: "Any human/shark interaction where injury occurs to a human, being alive at the time of the attack, the equipment worn or held is damaged (including surfboards and kayaks)

The Great White shark, Carcharadon carcharias. *With some justification, the most feared of all the sharks.*

or where imminent contact is averted by diversionary action by the victim or others." Criteria include provoked attacks, and attacks on divers feeding captive sharks, but data on incidents involving attacks on boats, dead bodies or animals are not included on the main file.

THE FACTS

Since the earliest recorded attack in Australia in 1791 (a fatal attack on a native female on the north coast of New South Wales) there have been 579 recorded cases in Australian waters which fit the above criteria. Of those, 182 (or 39 per cent) were fatal. Of the 579 victims of attack, well over 90 per cent, were males, so female readers can now rest more easily knowing that members of their gender are attacked on average only once every 5.9 years.

The tiger shark, Galeocerdo cuvieri. *A true oceanic garbage disposal unit, a habit which also makes it one of the recognised "man-eaters".*

Which part of Australia is safest from shark attack (apart from the obvious answer, Alice Springs)? The Shark Attack File indicates that, perhaps not surprisingly, New South Wales has the highest number of attacks and fatalities, closely followed by, again no surprise, Queensland. Victoria, on the other hand, has the second highest population, but a very low rate of attack. Tassie and the Northern Territory look pretty safe, and Western Australia, while recording double the number of attacks that occurred in Victoria, had the same small number of fatalities (seven). Lastly, while South Australia has long been regarded as the home of the Great White shark, only 42 attacks and 18 fatalities have been recorded in that state.

The species responsible for attacks are not often identified. Of all reported cases, the type of shark was not recorded in 389 attacks and 157 fatalities. This only leaves a small number to work with, but of the remaining 82 attacks, 23 were by whalers (eight fatal), 20 by whites (10 fatal), 19 by tigers (nine fatal), 13 by wobbegongs (none fatal), four by grey nurse (none fatal) and three by hammerheads (none fatal). Three of the grey nurse attacks were bites on divers feeding captive sharks, while the other was a provoked incident. In the case of the seemingly innocent wobbegong, however, most of the attacks were not provoked. In one incident, a wobbegong swam 6 metres off the bottom to latch onto the leg of a female diver. A final point here is that white sharks inflicted the highest fatality rate of 50 per cent, but then again, this means that 50 per cent of those bitten by an identified white have lived to tell the tale.

Another statistic that John West unearthed is the time of day at which attacks occurred. Of 73 cases which recorded time of attack, 46, or 63 per cent occurred

between 12 noon and 6 pm. Only eight (11 per cent) were recorded after 6 pm (two of which were fatal), whereas 19 (26 per cent) were recorded between dawn and midday. This certainly casts doubt on the generally held belief that dusk is a particularly dangerous time to be swimming.

The whaler, or requiem, sharks are also considered to be relatively dangerous due mainly to their coastal habit. Pictured is the bronze whaler, Carcharhinus brachyurus.

Despite the increased population and probable increase in numbers of people entering the sharks' domain, there has been a significant decline in the rate of shark related fatalities since the 1930s. This may be due in part to meshing of surf beaches (there have been no fatal attacks on a meshed beach since these programs were instituted), but could also be due to a decrease in either the relative or absolute abundance of sharks themselves.

RELATIVE RISKS

To put the chances of shark attack in Australia into perspective, John West has gathered some statistics on other causes of death from misadventure. Remembering that there have been 182 fatalities due to sharks recorded in Australian waters over 200 years of European habitation, it is telling to note that during the period 1968-91, 121 rock fishermen have been swept off the rocks and drowned and 37 surfboard riders drowned. During this same period, 32 shark encounters were recorded with only 1 fatality reported. Want more? There were 11 deaths caused by lightning from 1983 to 1986 (four killed by sharks during the same period) and, believe it or not, there are two to three deaths every year in Australia caused by bee stings.

The message from all of this seems pretty clear. Shark attacks are rare events. In fact, you have far less chance of being bitten by a shark than you do of winning the lottery; but don't forget to keep your head down in the next thunderstorm, and whatever you do, give that bee on the rose bush a wide berth. Perhaps when the general public and the media accept these facts we will be able to rise above the inevitable hysteria each summer and start thinking of and treating sharks as what they really are. Beautifully adapted apex predators that are integral and important parts of the ecology of the world's oceans.

Sizes of Sharks: Fact versus Fiction

A question I'm often asked, and one which is the subject of considerable debate is, how big do sharks of various species really grow? The acknowledged reference work for facts and figures on sharks is a publication of the Food and Agriculture Organization of the United Nations, entitled *Sharks of the World* by Leonard Compagno. This and other publications were referred to in compiling the following.

The largest of all sharks, and in fact by far the largest of all fishes, is the whale shark. Although growing to a reputed 18 metres, the generally accepted maximum measured length is actually 13.7 metres, with the most recent accurate maximum measurement being 12.1 metres. Amazingly, this behemoth feeds entirely by filtering plankton. The weight of such a beast would be very difficult to actually measure, but there is no shortage of estimates of weights in the literature ranging from 20 to 40 tonnes. (You will note here that the verifiable maximum size is a lot smaller than the reputed maximum, a common phenomenon as we shall see.)

Mako sharks, Isurus oxyrinchus, *don't grow as large as their close relative the Great White, although they are known to reach a respectable 4 metres and 500 kg.*

A relatively small tiger shark, Galeocerdo cuvieri; *a species which is reputed to grow to over 9 metres in length, although this is highly unlikely.*

The next largest shark is another filter feeder, the basking shark. This species is "credited at 12.2 to 15.2 metres, but most do not exceed 9.8 metres" – a rather cryptic way of saying that the largest verified measurement was 9.8 metres. Again, though, no hard proof of the higher figure of 15.2 metres.

Other large sharks' maximum lengths are given in the authoritative literature as follows: Tiger shark – reputed to be 9.1 metres, but can't be confirmed. Some females greater than 5.5 metres. One caught off Indo China in 1957 was apparently 7.4 metres, weighing 3110 kg. (Note the words "reputed" and "apparently".) Thresher – 5.49, possibly to 6.09 metres (bearing in mind that half the length of a thresher is its tail). Greenland shark – at least to 6.4 metres, and possibly up to 7.3 metres. Great hammerhead – 6.10 metres. Sixgill shark – at least to 4.82 metres. Mega-mouth – 4.49 metres (only a handful specimens are known of this deep water plankton feeder). Dusky whaler – 4.0 metres. Mako – 3.94 metres. Blue shark – 3.83 metres confirmed, but up to 6.5 metres has been mentioned. Grey nurse – 3.18 metres, and bronze whaler – 2.92 metres.

Scientists will only accept a length or a weight if there is proof that an actual measurement was taken. A mere mention in a report could easily mean that the figure is an estimate, and this is where problems emerge. If the shark has only been observed in the water, then any length must be an estimate, even if it has been compared with the length or beam of a boat. Even for sharks killed and measured with a tape, weights are also often quoted which have never been actually measured. The problem is compounded when an estimate (guess) in feet or pounds (always in neat whole numbers such as 25 feet, 5000 pounds) is converted into metric, to become in this case, 7.62 metres and 2272 kg. Naturally, the latter numbers sound very accurate, and therefore, are assumed by the reader to have been measured; hence a guess enters the "hard" literature as "fact".

WHITE SHARKS - THE FACTS

The maximum size of the white shark is undoubtedly the most hotly debated in the world of sharks. We all have an innate horror of the image of a huge predator; and the more huge, the greater the horror. It is not surprising then to hear stories or read articles mentioning white sharks of truly prodigious proportions. All of the problems mentioned above are compounded where white sharks are concerned, to the extent that some serious scientific investigations have been undertaken to try to separate reality from myth.

Dr John Randall of Hawaii published a paper in 1987 in which he refuted lengths, which were often cited in the literature, attributed to several large white sharks. The first was a famous set of jaws held in the British Museum of Natural History of a shark caught at Port Fairy Victoria (Australia) in 1870. The length attributed to the jaws was a whopping 36 feet 6 inches (11.1 metres). However, by measuring the jaws and the enamel on the teeth, Randall came up with a more realistic length of 5.4 metres (17 feet 8 inches).

A second shark caught in the Azores in 1978 was reported as being 29 feet 6 inches long (exactly 9.0 metres, by the way). However, lengthy investigations by Randall and others only resulted in a photo of the shark, which, by comparison with other objects in the photo, lowered the probable length to a more reasonable 20 feet or less.

Finally, a length of 6.4 metres (21 feet) for a white shark caught off Cuba in 1948 is often given as the "true" maximum for whites, since it was reported in the scientific literature. However, using scaled photographs of teeth and a vertebra of this shark, together with the known weight of its liver, Randall estimated the length of this shark to be no more than 5 metres (16 feet 4 inches).

The Great White shark is the largest predator on earth, but how big do they really grow?

So how big is the biggest white on record? The largest white shark, which has been completely verified, is one caught in a noose off Western Australia in 1984. It measured 19 feet 6 inches (5.94 metres) and weighed 1511 kg (3324 pounds). The heaviest measured weight, however, is greater: 1557 kg (3427 pounds) for one caught off Montauk, in the US, in 1986, although, interestingly, it only measured 17 feet (5.2 metres) in length.

The Weird, Wonderful Hammerheads

Some time ago, there was an intriguing television advertisement for a particular brand of screenwire door. The scene opens with a whaler shark chasing a scuba diver who just makes it behind the door before the shark crashes into it, and becomes – you guessed it – a dazed and bewildered hammerhead shark.

Most people would agree that the hammerhead is one of the most bizarre fish imaginable. Take a normal shark's body, and replace its head with a broad, flat flying-wedge, put the eyes and nostrils right out on the ends of the wedge, and you have the seriously weird hammerhead. Nothing else that swims has evolved a similar form, so, in the words of the late Professor Julius Sumner Miller, "Why is it so?"

Before attempting to answer that question, it is worth finding out a bit more about hammerheads in general. It might be a surprise to discover that there are nine species of hammerhead sharks worldwide, four of which call Australia home. Three of these are temperate species, which occur in game fish captures off southeastern Australia, but in unknown proportions. These are: the common or scalloped hammerhead (*Sphyrna lewini*), the smooth hammerhead (*Sphyrna zygaena*) and the great hammerhead (*Sphyrna mokarran*). These three are lumped together for record purposes, which is a pity because it is actually not difficult to identify each of the three species since each has a different shaped margin on the leading edge of the head.

Perhaps the first photo of a juvenile winghead shark, Eusphyra blochii, *caught in northern Australia, showing remarkable markings and colouration.*

This then brings me to the beautiful little shark shown at left. This amazing fish was captured in the Gulf of Carpentaria in northern Australia. The shark was actually caught in a creek flowing into the Gulf, and the angler who caught it was not only struck by the width of the head, but also by the juvenile shark's remarkable colouration – a virtual pelagic blue. I had not seen anything like this before, but could immediately identify it as the aptly named winghead shark. This species of hammerhead has evolved the most extreme extension of the head, to a width approaching half the length of its body. This is a tropical South East Asian species which is apparently not uncommon around northern Australia, however it is apparent from consulting my reference library that juveniles have not

A fish eye view of the seriously odd-looking hammerhead shark. Note the position of the eye and nostril. This is the smooth hammerhead, Sphyrna zygaena.

been especially noted. How else do you explain the absence of a description of the remarkable blue body and striking marks on the top of the "hammer"?

Hammerhead sharks have small mouths relative to their size, and are primarily confined to a fish and squid diet. However, very large hammerheads have been implicated in attacks on humans, and they are classified as one of the potential man-eating sharks. The three hammerheads listed all grow to a large size with, perhaps not surprisingly, the great hammerhead growing to the largest –- at least 5 metres – and therefore the most likely candidate for attacks on humans.

So why do hammerheads have such a bizarre modification to their heads? Scientific opinion leans towards a combination of likely reasons. Firstly, the flattened shape may act like a hydrofoil, increasing lift and manoeuvrability (bearing in mind that sharks are negatively buoyant and need to generate lift as they swim). I am unaware, however, of any tests of this theory in flume tanks. Secondly, the broad head spreads the eyes, making binocular (3D) vision more efficient. This seems reasonable, unless you've ever seen a hammerhead swim up a berley trail trying to find something to eat –- they seem to have an awfully hard time spotting a bait. Thirdly, the many sensory capabilities of the shark, such as smell (the nostrils are wide apart) and chemical, electrical and pressure senses, are increased by spreading them along the leading edge. To me, this seems the most likely function of the broad

A juvenile scalloped hammerhead shark, Sphyrna lewini.

head since, again through observation, hammerheads move their heads rapidly in sweeping arcs when hunting – not unlike biological metal detectors. I used to ponder on how such a bizarre body form had evolved, since I was only aware of the fully-fledged hammerheads which occur around Australia. Where was the missing link between normal sharks and these broad-headed wonders? Well, in fact, the missing links do exist in the Americas, where they are called bonnetheads, since the head is flattened into more of a rounded shape, without the marked lateral extensions. All in all, quite a remarkable group of fishes, I'm sure you'll agree.

The Cookie Monster

The marine world is indeed a fascinating and mysterious place. Strange creatures lurk in its depths and many have evolved bizarre means of making a living. This story begins with observations over many years of round, open, crater-like wounds on the sides of large marine animals. The wounds were always round and deep, about 4 to 6 cm in diameter, or were semi-circular cuts or round scars. Animals affected included tuna, swordfish, marlin, whales, dolphins and seals.

Well, marine scientists started to cogitate over the cause (as they tend to do) and came up with a wide array of theories. Over the years, these have included bacterial or parasitic infections, abrasions or, more commonly, the bite of some unknown animal. The most likely candidate to cause such a bite was the lamprey – a primitive, blood-sucking fish with no jaws which latches on to the prey like a leech and rasps through the skin and flesh with a circular, lined mouth. Lampreys were well known to early ichthyologists and it was natural that they should spring to mind. The only problem was that the wounds were much larger than could be accounted for by lampreys in general. They occurred in areas where lampreys did not inhabit, and the shape of the wound was not right (lampreys suck blood and leave a flat sore, but crater wounds are the result of a whole lump of flesh being removed as a neat plug). Also, crescent-shaped cuts and still-hanging round chunks of flesh suggested that something with very sharp teeth was at work.

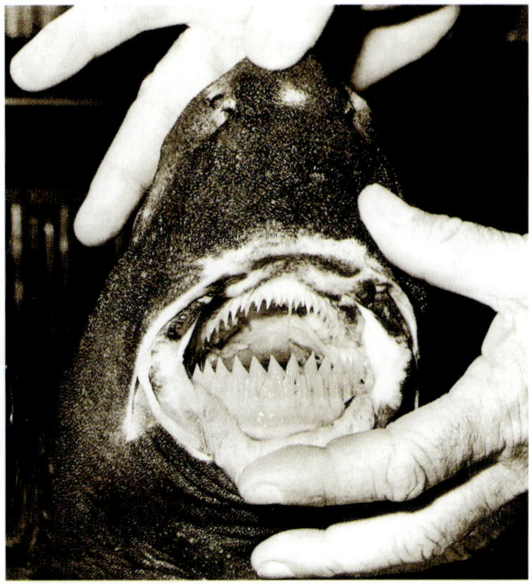

The business end of the ferociously "dentitioned" cookie cutter shark, Isistius brasiliensis.

So scientists started to look around for other candidates. One other possible culprit emerged from the ruck. It is a small shark in the dogfish family bearing the name *Isistius brasiliensis*. Known as the cigar shark, it only grows to a maximum length of 50 cm. It is caught, incidentally, in mid-water, and its distribution is similar to the distribution of animals found with fresh bites. It also has a particularly unusual set of teeth; those in the bottom row are sharp and flat and set closely together so that they look not unlike a band saw, while the top teeth are narrow and hook-like. The pieces were beginning to fall into place.

In 1969, a biologist named Everett Jones of the National Marine Fisheries Service in Honolulu was on a research cruise off Hawaii aboard the famous research vessel *Townsend Cromwell*. They were catching the occasional cigar shark and, out of curiosity, one of the research assistants held a specimen up to the side of a dead fish. The little shark made a biting motion and lo and behold, a crescent-shaped cut was made which, if it had been completed, would have resulted in a perfect crater wound

The ulcerated wound occasionally found on blue marlin, Makaira nigricans. *The hunt is still on for the culprit, but a suspect has been named.*

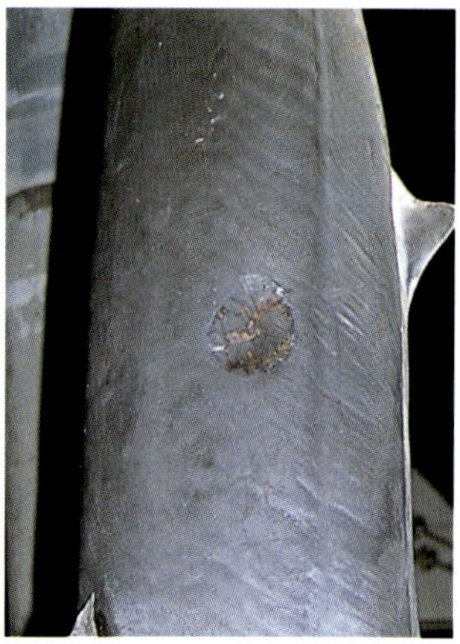

A healing cookie cutter wound on a mako shark. Bites on sharks are comparatively rare.

of about 4 cm diameter. The scientists then placed a second live cigar shark against another dead fish and, while this time it did not bite, it formed a strong point of suction with its mouth which enabled the fish to be picked up by holding the shark.

Finally, several fresh specimens of the shark were dissected and three were found to contain perfect plugs of tissue in their stomachs, two from large fish (based on the spacing of the muscle fibres) and one from a squid. Jones made the point in his publication of this account that the plugs would not be found in the stomachs of frozen or preserved specimens due to digestion of soft tissue continuing after death.

So with this series of simple but insightful observations, it seems that the culprit had been found. This has now been pretty well accepted, to the extent that the little parasitic fish has been dubbed the "cookie cutter" shark, for obvious reasons.

The next question is, of course, how does the cookie cutter achieve its aim of taking a bite out of much larger, usually predatory, fishes and marine mammals? The answer to this may lie in another feature of the shark's biology, which in fact gave it its Latin name, *Isistius*, after Isis, the Egyptian goddess of light. The cookie cutter shark is noted for giving off light at night. This was known as early as 1840 when a naturalist called Frederick Bennett wrote about two live specimens of *Isistius* which he had caught: "When the larger specimen was taken into a dark apartment, it afforded an extraordinary spectacle. The entire inferior surface of the body and head emitted a vivid and greenish phosphorescent gleam, imparting to the creature, by its own light, a truly ghastly and terrific appearance."

The light comes from little groups of cells called photophores arranged in patterns along the skin of the shark. This has led to speculation that the sharks' light pattern mimics that of deep-water squids and that they intentionally attract a predator towards them at night. (Cookie cutters apparently swim with squid schools and they certainly eat them, too, so maybe squid are also fooled). Many of the wounds are formed from the front of the predator, indicating that the larger fish or mammal has

The tell-tale crater-shaped wound left by the cookie cutter, this one on a broadbill swordfish, Xiphias gladius. *These wounds, both fresh and healing, are remarkably common on large pelagic fishes, as well as dolphins.*

been swimming towards (attacking?) the shark, only to have the tables turned. Perhaps they turn their lights off at the last moment (the shark is otherwise jet black), and make a dash for the flank of their would be attacker.

OTHER MARKS

Another odd wound has caught my eye from time to time. I noticed a number of blue marlin that came to the dock bearing rather nasty-looking ulcerated areas on their bodies. These lesions were remarkably consistent in appearance and in their position on the body, and provided two clues as to their possible cause. They were nearly always positioned just forward of the anal fin on the lower half of the body. Mostly there was one such mark, but sometimes two, symmetrically positioned on either side. The appearance of the wound was an oval shaped ulcer, often with horizontal ridges across the surface. In fact, it was not unlike the footprint of a tennis shoe. Unlike the case of the cookie cutter shark, we don't have a guilty party for this mark as yet, but I can suggest a pretty good suspect. The shape of the mark, and the "ripple-sole" effect, strongly suggest a pattern left behind by a remora, or sucker fish. The size is pretty consistent, indicating large remoras, probably around 60 cm or so in length (ie, not the small, white remoras which are often found living inside the gill chamber of billfish). It seems that this position, towards the rear of the marlin, may well be a point of low turbulence, so much so that the remora (if that is the true culprit) always takes up position in the same place. Again, I'm theorising, but the constant suction over one area of skin probably causes abrasion, cutting off of blood supply and ulceration. This explanation seems to fit the evidence, but as with many mysteries from the deep, it may be some time before proof comes to hand.

A Tale of a Whale (shark)

A lot of unusual marine phenomena go unnoticed, so it's nice to have a network of people out there who keep us up-to-date with the weird and wonderful things that happen from time to time. This is a story from a colleague in Papua New Guinea about a very small whale shark. It had been stranded in a tidal reef pool in Rabaul, and caused quite a bit of interest. Before discussing this amazing find, it will pay to outline what we know about whale sharks in general, even though that is not a great deal.

The whale shark is by far the largest fish in the world. The largest measured specimen was 12.1 metres long (39.7 feet) and was estimated to weigh just over 12 tonnes, but it is likely that the maximum size is larger.

The whale shark is a single, worldwide, tropical species that is relatively common, even though its conservation status is under review. It is a harmless filter feeder, although it does not passively strain its food by moving slowly through the water like some of the baleen whales (and the other filter feeding shark, the basking shark). Rather, it is a suction feeder, often feeding in a vertical position with its head towards the surface, suddenly sucking in large volumes of water containing its food items. Usually their food consists of small fishes and crustaceans, but they have also been recorded with larger fish such as tuna and remoras in their stomachs – perhaps inhaled accidentally rather than by design.

A rarely encountered juvenile whale shark, Rhincodon typus, *accidentally trapped on the outgoing tide in Papua New Guinea.*

The finding of this small PNG specimen raises questions on the reproduction of whale sharks. For many years, it was thought that the species laid eggs. This theory was entirely due to a single egg containing a whale shark embryo trawled up from 31 fathoms off Mexico in 1953. The egg measured 30 x 14 x 9 cm, and the embryo inside was 36 cm long, apparently near full term. However, the papery egg case was very thin, and did not possess tendrils like other shark eggs, possibly indicating that the eggs are normally internal, and that this egg had been prematurely aborted. The mystery appears to have been solved more recently when a pregnant female was dissected and found to

The largest fish in the ocean, the whale shark. Its biology is still largely a mystery.

contain over 300 pups which were in the process of hatching from eggs while inside the mother. With this single observation, the whale shark has been demonstrated as not only the largest fish, but also the shark that produces the largest litters of pups.

Between birth and the large adults usually encountered, there is a large gap in observations of juveniles. The smallest free-living whale sharks found so far were 55-56 cm long, so the PNG specimen was a fair bit bigger (roughly about 120 cm). Naturally, the big question is how old would this particular baby whale shark be? Knowing that they are born at about 40 cm, this one is certainly no newborn, but it is also likely that the species grows quite quickly. Perhaps this youngster was still in its first or second year, but of course there is no way of knowing with any certainty.

In recent years, the people at CSIRO have been conducting some very interesting research on whale sharks on both the east and west coasts of Australia. One of the most famous places in the world where whale sharks congregate, seemingly on an annual basis and where they can be observed in very clear water, is Ningaloo Reef off Exmouth, Western Australia. Here, Dr John Stevens (of CSIRO), and others, have

Whale sharks often swim close to boats, and even rub up against them. It is thought they do this to clean off parasites, presumably mistaking the boat for a floating log.

been attaching "smart" tags to these ocean giants in the hope of recording their travels throughout a whole year, and then retrieving the data-laden tag if and when the shark returns on cue the next time round. At the time of writing, they are awaiting their first tag retrieval with great interest. Knowledge of whale shark movements and stock structure in the Indian Ocean is particularly important since they are actually fished commercially off India and other countries.

Off eastern Australia, whale sharks occur with considerable regularity in the Coral Sea off the Great Barrier Reef. There, they are often associated with seasonal aggregations of lantern fishes, on which they feed, and which also attract large numbers of bigeye and yellowfin tuna. In this area, whale sharks seem to act as natural FADs (Fish Attracting Devices) and it is not uncommon to see whole schools of bigeye tuna stacked vertically under each whale shark. The object of the CSIRO research in this area is to attach satellite pop-up tags to whale sharks to not only plot their movements but also possibly the schools of bigeye tuna that swim with them.

The whale shark is certainly an impressive animal – a true example of charismatic mega-fauna if ever there was one – and an illustration of how little we really know about even the largest denizens of the deep.

The Mysterious Mega-mouth

New information regarding another giant filter-feeding shark – the mega-mouth – always makes fascinating reading. Its existence has only been known about for the past 25 years or so. The enigmatic species of shark, the aptly named mega-mouth shark, was only known from two specimens, making it a definite candidate for the title of world's rarest shark. Those two specimens were rather famous – the first, 4.46 metres long, being found tangled in the lines of a research vessel off Hawaii in 1976 and the other, just 3 cm longer, also tangled in fishing gear off Catalina Island, California, in 1984.

It was with some surprise, therefore, that I was (politely) informed by a marine science student that this information was way out of date, and that there have in fact been at least 14 mega-mouth sharks recorded so far. And where did this new information come from? The Internet, of course. Sure enough, there on a site specifically designed for the purpose was a list of all the known mega-mouth specimens, apparently compiled by a reputable source – a Dr John Morrissey of Hofstra University in New York (you can find it at www.flmnh.ufl.edu/fish/Sharks/Megamouth/tablemega.htm).

One of only 14 known specimens of the mega-mouth shark – this one was stranded on the coast of Western Australia.

The list is particularly interesting since it gives the locations of the capture or stranding of each specimen, its size and what happened to it. While most of the sharks ended up in museums, six of them had been either released alive or discarded, so there may be some problems with the certainty of their identity. One on the discarded list was quite small (less than 2 metres) and apparently recorded from a remote location (Senegal). Unless photos exist of this specimen, it may be a doubtful record. Nevertheless, assuming that all are properly identified and confirmed mega-mouths (as indicated on the website), the list indicates that specimens have now been found in all three major oceans. In the Pacific, off California (three), Hawaii (one), Japan (five), the Philippines (one) and Indonesia (one). In the Indian Ocean, one specimen beached at Mandurah, Western Australia, (in 1988) and in the Atlantic Ocean, one off southern Brazil, and one off Senegal.

Even with the extreme rareness of such finds, believe it or not there has actually been a book produced on the biology of the mega-mouth shark. This is mainly based on one specimen that was recovered in good condition in Japan in November 1994. A team of experts was invited to study the carcass and, as a result, an amazing 21 research papers were written and compiled into a full volume.

Unfortunately, I have not yet been able to obtain a copy of this rare work, but the contents indicate that many of the papers present data obtained from examining the soft parts and general anatomy of the shark. Several scientists also took the opportunity to extract DNA and compare it with other sharks, and the results were fascinating. The first descriptions of the mega-mouth from Hawaii placed it in its own unique family, and the DNA analysis subsequently confirmed this. It showed that the mega-mouth was completely different from the superficially similar filter-feeding basking shark, meaning that all three of the giant filter-feeding sharks, the whale shark, basking shark and mega-mouth, evolved independently of each other, almost certainly from separate non-filter-feeding species in the first place.

Summarising what is now known about the mysterious mega-mouth, it is presumably a deep dwelling, slow moving animal since its skeleton is weak for its size and it has relatively low muscle mass and soft fins. Stomach contents indicate a diet of mainly euphasid shrimps, which ascend and descend between depths of 300 and 1100 metres during the day, rising to within 75 metres of the surface at night. Other material found in its stomach suggests that it also eats jelly fish. The confirmed specimens have mainly been measured at between 4.5 and 5.5 metres, and both sexes have been recorded. Interestingly, the mega-mouth is the only species of shark or ray known to continually suffer attacks from the cookie cutter shark. This is thought to be more due to their soft skin rather than sluggishness.

The recently discovered mega-mouth is a deep water filter-feeder which engulfs its prey as it swims.

It was surmised by world shark expert Leonard Compagno when the first two specimens were found, one in Hawaii, and one in California, that this was very likely a widespread species, like the other two filter feeding sharks, roaming over a broad area of the tropical and sub-tropical Pacific Ocean. So far, as the list above shows, these predictions have proven to be spot-on, with the range of the mega-mouth now seemingly worldwide.

One of the big questions remaining to be answered about the mega-mouth is, are they really rare, or just not encountered very often? I would tend to opt for the latter because of their obvious widespread distribution, and because the chances of them being caught or seen are so slim, due to their unusual habitat. In any case, that such a large shark could exist undetected until the last quarter of the 20th century illustrates well how much we still have to learn about life in our oceans.

Conservation and Fishing

Humans have fished since time immemorial, but in the twenty-first century, is fishing still compatible with the ideals of conservation? Can fish populations withstand the pressures of commercial and recreational fishing, let alone the threats from other human activities which destroy habitat and pollute their environments? We will always go down to the sea to fish, and with care for the environment, and the fish themselves, such simple pleasures will surely be passed on to those who follow.

Clean seas and good management equal healthy fish stocks

Do Fish Survive Catch and Release?

These days, as more and more anglers become conservation oriented, catch-and-release and tag-and-release are part and parcel of fishing in Australia. The huge support received by organised tagging programs shows just how ethical fishing has become. Size limits on most species also mean that many undersized fish are released, and, more recently, fisheries managers in several Australian states have introduced a wide range of bag limits, again leading to more and more fish being released after being hooked. But the questions which is being asked is, how many released fish survive? What can be done to maximise survival? Is tag-and-release or clean release a waste of time, or are we really being effective in trying to conserve our limited fish stocks?

An important overview of the scientific literature on this subject sheds considerable light on survival of fish after hooking. Titled "Hooking mortality: a review for recreational fisheries", and published in Volume 2 of *Reviews in Fisheries Science*, two Texan biologists, Maurice Muoneke and Michael Childress, amassed details of some 132 studies which were relevant to hooking mortality, and summarised them in a number of ways. It is important to realise that most of the studies were conducted in North America on salmonoids (trout and salmon), with smaller numbers on other freshwater fishes such as catfish, pike, bass, crappies and walleye. Very few studies were found for saltwater fish, and most of those were confined to members of the striped bass family and the drum family (the latter contains our own jewfish or mulloway). Most of the studies have been carried out by catching fish on angling gear, confining them for a period of time, and measuring the resultant mortality, although recapture rates of tagged fish are also important sources of information. Sometimes

The growing trend towards catch-and-release means that we need more research into survival of released fish. The author about to release a small barramundi, Lates calcarifer.

fish have been caught by anglers under normal fishing conditions, sometimes by scientists using carefully replicated techniques. Numbers of fish per experiment are very variable, and the lengths of the experiments in terms of days also have varied greatly. Nevertheless, in summarising this large amount of information, some generalisations seem to be able to be stated.

Careful handling of fish is important prior to release.

Combining all results, 41 per cent of all studies showed mortality due to hooking to be 5 per cent or less, 52 per cent showed mortality at 10 per cent or less and 70 per cent of studies showed hooking mortality at less than 20 per cent. These figures varied a little amongst types of fish, but in general they were consistent enough for us to be pretty confident that those sorts of mortality rates are the ballpark figures in most situations. Some of the variables affecting mortality due to hooking and releasing were covered, too. A few studies have looked at effects of catching fish from different depths and have suggested that deflating the swim bladder may indeed increase survival. On the other hand, trying to deflate popped eyes by inserting a sharp object into the eye cavity of the skull resulted in greater mortality than non-deflation (ouch!).

While a lot of the studies quoted only lasted for a few days after capture, a reasonable number of experiments conducted over longer periods suggest that most of the mortality happens within 24 hours of hooking and release, and that mortality is substantially complete after about six days. (I would like to see more of this work done, but long-term experiments are very costly.)

Should we use single or treble hooks, barbed or barbless hooks if release is intended? Perhaps not surprisingly, barbless hooks have usually been shown to cause less mortality than barbed ones, but not in all cases. On the other hand, it may come as some surprise to find that treble hooks are not as bad as singles. Trebles tend to cause less deep-tissue damage (fish usually can't swallow them) but this may sometimes be counteracted by the longer handling times in unhooking fish caught on trebles.

Concern has sometimes been shown for the non-lethal effects of catch-and-release, such as those on growth rate or reproduction. Most of the studies on trout show that reproduction is not affected by prior hooking, but one Australian study on snapper was quoted as indicating that hooking-related stress may impair their reproductive

Holding a small billfish vertically is probably okay, but it is not recommended for large fish, which might damage internal organs.

development, although this effect may have been due to the combined effects of angling stress and subsequent confinement.

Quite a lot of studies strongly suggest that removing hooks from deeply hooked fish is not a good idea. This tends to tear vital organs and increases mortality, so the advice if a hook has been swallowed is to simply cut the line or leader. Captive experiments have shown that many fish have a remarkable ability to shed hooks, some appearing to actually "cough" the hook out.

Unfortunately, the review of hooking studies shed little light on the issues of live bait versus dead bait versus lures, and the stainless versus galvanised hook debates, mainly because few, if any studies have yet been done to try to answer the difficult questions involved. There is still a lot of work to be done on hooking mortality, but this summary is a good starting point. As time goes on, more and more captive, and tagging studies will be completed, providing anglers with a much clearer understanding of what happens to their catch after releasing it to the wild blue yonder.

Survival of the Fittest?

I think just about every angler I know has at least one injured fish story. You know: the tailor with a mangled jaw, the yellowfin tuna with a chunk missing from its flank, the bream with the deformed backbone, and so on. The common element in the telling of these stories is amazement at the fact that the fish was alive at all, let alone that it was quite capable of taking a bait or lure and also capable of putting up a more than respectable fight. Take the case I know of a jewfish caught on the Gold Coast Seaway. The fish's eye socket was very distinctive – it in fact had no right eye at all – and its lower jaw was lopsided, but it was in good condition and had no trouble feeding. As to the cause of the deformities, we can only speculate, of course. However, it would be a fair bet to assume that fish like this have, at some time in the past, come off second best after an encounter with a baited hook.

Eye injuries are pretty common when a fish is hooked in the roof of the mouth, and I've seen my fair share of tailor missing one eye. Broken or dislocated jaws are also seen reasonably often, and again, a hook is the most likely explanation, although some of these and other injuries could, of course, quite easily be the result of encounters of the non-human kind. Things that come to mind are injuries caused by predators, by spiky or aggressive prey, or in the case of injuries to marlin bills (again, quite common), the result of collisions with floating objects while chasing sheltering bait-fish. The other category of affected fish – creatures which may still be capable of a relatively normal lifestyle – are those with deformities present from birth, or at least caused at a very early stage in the fish's life cycle. These sorts of deformities can be

A small yellowtail kingfish, Seriola lalandi, *with a big problem. Most likely, the lure will fall out, leaving little or no permanent injury.*

caused by genetic defects, early disease, parasite infestations, shock (such as from Naval depth charges) or toxic substances (known as teratogens, or "monster makers".)

I have often been asked whether such injured fish, because of their handicap, need to be more aggressive just to keep up, just to eke out a living in their watery fish-eat-fish world. Unfortunately, we don't really know. Intuitively, the chances are that fewer injured or deformed fish make it compared with healthy fish. But since we do see plenty of examples of survivors swimming in the face of adversity, perhaps there may well be some basis for the idea that they do struggle harder to survive alongside the fittest members of their ilk.

An amazing case of recovery from injury. A blue marlin, Makaira nigricans, *showing a completely healed gill arch which had been torn during an earlier incident. This would be the equivalent of surviving a severed carotid artery.*

I have only come across one short scientific paper relevant to this topic, based on observations by a Japanese colleague of mine, Izumi Nakamura, of a deformed bluefin tuna he found in his local fish market one morning. Landed with other members of a school of fish of about 10 kg, this particular specimen was pug-headed – its top jaw was poorly formed – and its bottom jaw protruded considerably. In fact, it had the fish equivalent of a harelip. This condition is not unknown in tuna and other species such as trout, and it is generally agreed that it is a congenital deformity, that is, it occurs because of incomplete development of the skull. Anyway, Dr Nakamura carefully measured the fish as well as its normal companions and came to some interesting conclusions. The pug-headed fish was, in fact, lighter than the other fish in the school, even though it was the same age. This was interpreted as indicating that, because being part of a school is so important for small tuna, the little fellow had to exert a little more energy to keep up with its more streamlined companions, and probably wasn't as efficient at catching prey either.

Over the years, I have seen and measured many hundreds of marlin, of which, quite a few had missing bills, missing or injured fins, or had obviously recovered from horrendous flesh wounds, either from shark attack or from gaffing misadventures. While some of these fish were in poorer condition than would be expected for their length, most were quite robust, and apparently normal in every other way. And as far as doggedness or fighting qualities were concerned, in just about every case, the anglers who caught these fish have assured me that they fought just as well as, if not better than, normal specimens. So, even if we can't carry out the perfect controlled experiments to answer these sorts of questions with conviction, most of us would agree that fish can and do survive some pretty shocking injuries and, having done so, apparently can hold their own alongside their fitter companions.

An amazing case of a marlin whose bill had been completely broken off at some stage in the past, yet was apparently none the worse for wear.

What is a Normal Fishing Year?

It seems a very human trait to think of the seasons and nature as fitting into a comfortable, annual cycle of events. Seasons come and go in sequence, flowers bloom, birds and bees do their thing, and all is as it should be in the great chronological machine of creation. With that sort of mindset, we tend to look for order, for repetition and prediction about natural events. And so with fishing. We grow up with expectations that fish too follow set seasonal patterns, and, yes, sometimes they do. But of course, often, they don't. And they don't because of the inherent variability of the environment in which they live.

To gain some idea of the immense potential for variability within marine ecosystems, consider the incredible fecundity of most oceanic fish, that is, the number of eggs they produce. A mature female tailor, flathead or bream may liberate over a million eggs in a spawning session, while a giant bluefin tuna or marlin will produce as many as 100 million or even 200 million eggs at a time. Skipjack (striped) tuna spawn several hundred thousand eggs every day or so for months on end. In all of these cases the eggs themselves are no more than a millimetre in diameter. They are fertilised externally, and float in mid-water or on the surface for a few days before hatching into tiny larvae. Images of the ocean filled with a writhing mass of larvae come to mind but, obviously, the mortality of eggs and larvae must be massive in these early stages of life, and this is a crucial aspect in studying and understanding the dynamics of fish populations and their fisheries. The numbers of surviving juvenile fishes at any time will fluctuate depending on a whole range of factors such as spawning success, food availability for the larvae, predators (a major factor), weather (such as El Nino), and success of settlement of larvae into suitably protective and nourishing habitat. All of these virtually unpredictable variables mean that, even if human beings had never existed on this planet, the numbers of fish in a given population would fluctuate over wide ranges, in an apparently random fashion.

Predicting good and bad fishing years is harder than predicting the weather. That's probably because they both have a lot in common.

Observing fluctuations in fishing success from season to season can give some insight into the variability of marine systems.

Scientists have tried to grapple with these problems for many years in their attempts to predict fish numbers and, therefore, to manage catches. In the US, future striped bass catches are predicted on the basis of the abundance of very small juveniles in their first year, but what are the main causes of larval and post larval survival or mortality? As mentioned, factors such as temperature, rainfall, prevailing currents and plankton concentrations all can and do play a part, often interacting in complex ways. Other less obvious environmental conditions may also come into play. One such factor may well be wind, and scientists at the CSIRO Division of Fisheries laboratories in Hobart have been investigating possible effects of prevailing winds on recruitment (ie survival) of larval fish. By plotting the abundance of various species such as gemfish each year against the number of days of strong west winds, they have found a strong correlation between the two. Periods of strong winds seem to be also correlated with good years for survival of juvenile scallops and lobsters in Tasmania, leading to a general conclusion that this is a widespread effect. While the correlation is strong, the scientists are yet to come up with any proof of how the wind might affect recruitment. It is possible that the winds carry the surface floating eggs and larvae to favourable locations, but it is also possible that whatever causes the winds to be strong in a particular year might also affect recruitment in some way. Whatever the mechanism, fisheries managers would still be able to use the fact to better predict availability of fish in the future.

One of the problems with these types of studies in which environmental variables are matched with strong and weak "year classes" of fish is that the data usually only go back in time for 20 or 30 years, sometimes a lot less. These time frames simply may not be long enough to explain the whole picture in many cases, especially for long-lived species.

My own observations of the changes in the distribution and abundance of pelagic fish off eastern Australia only span a little more than two decades. I am still not sure what constitutes a so-called "normal" year, it looks like I may have to be around for many more decades before I can answer that question.

A mature female tailor, flathead or bream may liberate over a million eggs in a spawning session, while a giant bluefin tuna or marlin will produce as many as 100 million or even 200 million eggs at a time.

Fish Sauce

Like many Australian kitchens these days, mine is stocked with a multitude of multicultural ingredients. Asian food, especially Thai, rates very highly, and my cupboards bulge with all sorts of exotic sauces, spices and condiments. Most trips to the supermarket result in an addition or two to the collection, and last time I scored a particularly interesting bottle of fish sauce from Thailand. To the uninitiated, fish sauce is to southeast Asian cuisine what salt is to European or Australian food or what soy sauce is to Chinese. It is a universal ingredient of most dishes, and a bottle is always on the table to add a few drops to the served food. Anyone who has ever been to a Thai restaurant would have consumed some fish sauce in one form or another.

My latest prize was labelled: "Fish Sauce: First Press Virgin, Aged Over One Year". Pretty impressive, although I'm not entirely sure about the virgin bit, and I wasn't aware that fish sauce improves with age. In any case, this particular label brought back some memories and thoughts about the product, and about the use of fisheries resources in different countries.

Several years ago, I had the opportunity of working in Vietnam on the development of an economic plan for the central Province of Quang Nam-Da Nang. My role was to research the current and potential use of the marine resources of the area, which involved extensive travel and inspections of all aspects of fish and fisheries production (not to mention eating lots of the produce). One of the ports of call on the itinerary was a fish sauce factory – sounded interesting, but I was unaware of what awaited me. I don't know if you've ever had the experience of walking into an invisible olfactory wall, but this one stopped me dead in my tracks as I tried to find clear air. After suppressing the urge to bring up the previous night's dinner, and after acclimatising to the alien atmosphere (presumably my olfactory nerves had been fried), I politely went on with the inspection.

Australian Anchovy
Engraulis australis
12 cm
0.01

Anchovies are widespread throughout the world. This species occurs right around the southern half of Australia, as well as offshore. There are several species through southeast Asia that are important components of fish sauce.

Row upon row of open vats containing fermenting fish in various stages of decomposition, not that the fish themselves were recognisable as such, awaited me. Fish sauce is basically the liquid left after fish have been allowed to ferment to completion. The manager of the factory was very proud of his product, and pointed out that there were many grades of sauce depending on such factors as species of fish used, the process and so on. (Hence the premium variety mentioned above.)

The liquid is remarkably clear, the colour of cold tea. The smell of the sauce is certainly fishy, but not unpleasant, and when cooked or added to food, adds flavour and saltiness.

So why, apart from having something to do with fish, would I be going on about fish sauce to this extent? Well, during my tour of the factory, I asked a few questions about the types and quantities of fish used in the production of fish sauce, and received a few surprises. The fish used can vary, but are nearly always very small – no more than a few centimetres long. Some of these are species such as anchovies and herring, which do not grow large, but many others are juveniles of species that grow to larger sizes as adults.

I was told that it takes between 1.5 and 2 kg of fish to produce a litre of sauce. Now, the production of fish sauce within this province, for internal consumption, was a hefty 5 million litres, and growing, meaning that a staggering 7 500 to 10 000 tonnes of small fish was being used annually in this one province for fish sauce production. The total production of fish sauce in Vietnam is at least 10 times this amount, needing a huge quantity of fish biomass. Obviously this raises a number of important questions, not just in Vietnam, but also in other southeast Asian countries

Pony fish
(toothpony)
Gazza minuta
14 cm

The pony fishes all belong to the family, Leiognathida, which is a large, mainly tropical family of small, schooling fishes. Some species grow to as large as 25 cm, but most are caught in southeast Asia at a very small size.

A Vietnam market scene. Nothing is wasted, and small fish such as those in the foreground are used in the manufacture of fish sauce on a large scale.

where fish sauce is a staple part of the diet. After all, this quantity of fish is not being turned into protein, but into a condiment.

I would make the point strongly here that I'm not being critical of the historic cultural mores which exist in southeast Asia. Cultural norms are difficult if not impossible to change, and fish sauce has been part of the culture in these countries for hundreds of generations. However, with the economies of the region expanding at an enormous rate, and fishery technology keeping pace, there is a real danger of overfishing on a vast scale. And of course, the problem doesn't stop with fish sauce. There are other products such as Blachan, a potent paste made from small prawns which is universally used for example in Indonesia by its 100 million plus population.

As a last example (and there are many others), did you know that the quantity of prawns imported annually into Australia (about 9 000 tonnes) is not much less than the quantity exported (about 11 000 tonnes)? The trouble is that the imported prawns tend to be small, hand-shelled in Asian sweatshops, the ones you often get on a sandwich, or on seafood pizzas, etc, while we export the jumbo prawns from the lucrative Gulf of Carpentaria fishery.

It's difficult to know what the solution is to these sorts of imbalances, other than to hope that the fisheries managers in these regions perceive the problems and act before it is too late.

Flotsam and Jetsam

A gentle stroll along a beach must surely be one of life's most simple pleasures. Not only is it healthy and invigorating, but if you keep your eyes open, it can be interesting and educational as well. The art of beachcombing is the art of observation, and it's amazing what you can find on any beach if you take the time to look.

Like most people, I usually walk along the water's edge (the sand's firmer), occasionally picking up the odd freshly cast-up object. But if you take the time to wander up the beach, to the high-tide line of debris, that's where most of the really interesting stuff, the "flotsam and jetsam" may be found. But what does this term mean? Well, flotsam really means anything that floats, while jetsam literally means anything jettisoned, or literally thrown overboard. Nowadays, we take the words to mean anything washed up on a beach, either natural or man-made.

As an exercise in beachcombing, I strolled along Kurnell Beach on the southern shores of Botany Bay in Sydney. Armed with a bucket, it didn't take long to fill it with the good, the bad and the ugly.

The Good

Seaweed, driftwood and shells are signs of a healthy environment.

The natural or biological component of beach debris is an indicator of the health of the adjacent waterway. In my little sample, there was plenty of seagrass, kelp and other algae, as well as a large number of sea urchins (which graze on seagrass), sponges, and of course, shells of a wide variety of molluscs including clams, scallops and sea snails. There were also numerous small cuttlefish shells (the internal skeletons of cuttlefish, which seem to be on every beach) as well as various crab remains and even the carapace of a spiny crayfish. Not a bad haul for a short stretch of urban beach, and a fair indicator of a reasonably healthy waterway. But there were plenty of examples of not so good flotsam and jetsam too.

The Bad and the Ugly

Unfortunately, and inevitably, this particular beach was littered with all sorts of plastic debris. As we all know, plastic takes many years to break down, and is often ingested by a wide range of marine animals including fish, birds, turtles, dolphins and whales. This is bad enough, however there is one category of man-made jetsam that causes immense damage, and is also very close to home as far as recreational fishing is concerned. I'm talking about fishing line. In my short walk along this one beach, I found bunch after bunch of tangled line, not to mention short lengths with terminal tackle still attached. Now admittedly this was on an urban beach where one would expect a lot of fishing, but nevertheless, multiply this by a suitable factor for the coastline of Australia, and we can begin to see the potential magnitude of the problem.

Beachcombing can tell us a lot about our marine environment, both good and bad. Biological debris indicates the health of a waterway, but plastics and synthetics, especially discarded fishing line, can cause serious damage to wildlife.

One of the downsides of discarded or lost fishing line is entanglement of seabirds. I've long been aware of this problem through reading overseas studies, and have also seen the occasional entangled seagull. However, I was unaware of the scale of what was really going on under our noses until I was contacted by Lance Ferris of Australian Seabird Rescue. This group spends many hours in unpaid work saving the lives of many, many seabirds, a high proportion of which are tangled in fishing line. Lance tells me that his group, which only operates on the northern New South Wales coast, has rescued well over 300 pelicans that were entangled in line, either with or without hooks.

Conservatively expanding this to the New South Wales coast, he estimates that at least 5 000 pelicans per year would become entangled. Pelicans are apparently especially prone to this problem since they spend a lot of their time poking around popular fishing areas. One observation is that set lines appear to be particularly dangerous to pelicans.

As well as pelicans, Australian Seabird Rescue has recorded a staggering 52 species of seabirds entangled with fishing line, and again, that's just on the north coast of

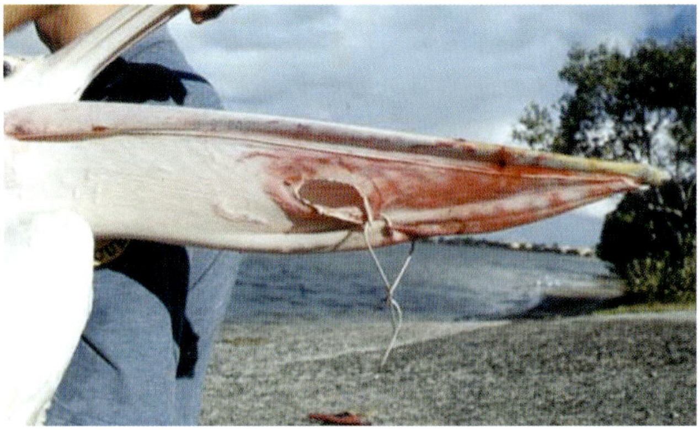

"Out of sight, out of mind": discarded fishing line and hooks cause much unnecessary stress to wildlife, especially seabirds like pelicans.

New South Wales. And it's not only adult birds that tangle and die. Many species of coastal and even inland birds use fishing line in their nests. The relatively rare Australian osprey is particularly fond of collecting clumps of line for its nests, with the macabre outcome that their chicks are often strangled as a result.

The problem with this issue is that it is largely unseen. I would like to believe that no angler would knowingly tangle a seabird and leave it to die, but here we have a classic case of "out of sight, out of mind".

In some ways, the issue of fishing line and seabirds is a little like that of accidental road kills of nocturnal mammals. You can't ban cars, but if you think laterally, you might be able to lessen the impact (like providing under-road tunnels for the mammals). In the case of fishing line, one obvious action is to keep ramming the message home: never throw line away. Take it home and dispose of it properly. And it goes without saying that if you come across discarded line, or pull up a snag or weed with fishing line attached, remove the line, don't throw it back. Never forget, you as an individual can make a difference.

Is Fishing OK?

A very large number of Australians would probably consider fishing to be a right, rather than a privilege. But there are others who, if they had their way, would ban fishing tomorrow. The main anti-fishing push comes from the animal welfare lobby and the arguments against fishing take many forms, and seem to shift ground when necessary. In the past, the anti-fishing lobby has been marginal and has not been taken very seriously. However, a couple of recent, well-publicised captures of record game fish have given the movement a relatively soft target, and an opportunity to woo the non-fishing public with its arguments.

The main, over-riding argument used against fishing is simply that it is cruel. Furthermore, catch-and-release (or tag-and-release) fishing is regarded as unacceptable since the aim of this type of fishing is to have fun at the expense of another animal, and worst of all, catching fish for records is seen as simply an ego trip. In the latter two cases, catching fish but not eating them is a particular anathema to anti-fishing groups, although they would still oppose recreational fishing for food. These arguments sound reasonable, and certainly seem to push the right buttons with some sectors of the media and the public. But how legitimate are they?

The whole realm of the use of animals by humans forms a fairly significant branch of philosophy, and can lead down some tortuous paths. One of the most coherent summaries I have come across in this area of debate is in an essay by Michael LaChat, a Professor of Christian Ethics at The Methodist Theological School in

To those who practise the art, fishing is a peaceful and rewarding form of relaxation. But some would argue that fishing is not entirely acceptable.

The ethics of fishing might be argued ad infinitum, but there is no denying the joys it can bring to the human condition.

Ohio. Entitled "An argument in Defence of Fishing", LaChat points out that many of the anti-fishing proponents tend to greatly exaggerate the similarity between animals and humans, thereby giving their arguments apparent greater credibility than they deserve. For example, an oft-quoted study of "pain" in fishes in 1994 equated escape behaviour with pain perception. However, as LaChat noted, even brain-dead humans may show reflex escape from pain stimuli. This is a key concept, in that "the human body can 'perceive' and even react to 'pain' stimulation without being aware (or conscious) of it. It is quite possible that this is the norm in fish, so the question is: is non-conscious 'pain-perception' worthy of moral consideration?" (In other words, if a fish is not conscious of pain, as we are, should we be concerned from a human point of view?) LaChat answers: "I think not. I don't even think the terms 'pain' or 'perception' should be used in such cases, since both imply conscious states that are enabled by a neocortical substrate." The neocortex is the part of the brain in humans which makes consciousness and experience possible, and there is little if any evidence that the neocortex exists in the brains of fish. LaChat concludes, "Therefore, it is unlikely that fish consciously 'experience' so-called 'pain' stimuli at all." He concludes this part of his essay with a strong and clear statement: "What we might give to [anti-fishing writers] is the agreement that the more closely an animal's brain resembles our own, the more careful or concerned we ought to be. Beyond that, anglers ought to concede them nothing."

In looking at the moral aspects of fishing, LaChat also makes the point that the benefits of fishing may well outweigh the costs to individual fish, even if we give them the benefit of sentience and, therefore, pain perception as we know it. Such well-known and well-documented human benefits can be psychological, spiritual, nutritional and economic. Furthermore, because anglers have an incentive to ensure that fishing continues into the future, they can (and do) help to protect healthy fish populations and the habitats on which fish depend.

Many of the anti-fishing lobby maintain a purist, non-intervention stance for humans on the planet. That's okay, but are they always consistent? Do they live in houses that now stand where natural wilderness once flourished? Do they eat the product of any living thing, even if they are vegetarian? Are we humans now, having been an integral part of the environment for millions of years, suddenly to become bystanders and observers, looking but not touching? Surely if fisheries management and fisheries science are legitimate disciplines, a sustainable resource should be able to be fished and enjoyed, with the community at large benefiting from all of the above benefits and spin-offs of these activities.

Animal rights activists may argue against fishing, but are their arguments based on fact?

LaChat puts it this way: "Life negates life, and if we have an obligation to future generations for ensuring biodiversity and ecological well-being, then surely we ought to be active managers of fisheries too [and not stand passively by]. We are part of the natural order. By omission or commission, we are predators as well as conservers."

Once Bitten, Twice Shy?

A topic that has fascinated me for as long as I've been involved with fishing is that of hook avoidance. Can and do fish learn to avoid hooks, becoming hook-shy, and if so, how does this happen?

A number of observations, as well as some hard science point to the reality of hook avoidance in fish. For example, the history of open ocean long-lining shows a well documented, standard pattern. When long-lining starts up in a "virgin" area never before exposed to long-lining, catch rates are often extremely high, but fall dramatically within only a few years, even though the overall catch may still be relatively low compared with the estimated fish population. It is often argued that this phenomenon may be due to a virgin population of fish containing a high proportion of genetically naive fish, that is, fish that are simply more likely to get caught. The new fishing method initially creams off a lot of these naive fish, leaving behind the more cautious individuals, which pass on those genes to their offspring.

Another example of naive fish in a population being more susceptible to hooking is that of reef fish. In a series of experiments in the Great Barrier Reef Marine Park, some long-term protected reefs have been opened to fishing to see what would happen. Immediately after opening the reefs to fishing, catch rates of coral trout and other reef species are initially very high, but only for a short time, reverting to normal catch rates similar to those of exploited reefs very quickly. Divers going below the surface to see what is really going on report that the newly opened reefs still have healthy populations of fish, but that after the initial surge of fishing, the remaining fish seem to be far less prone to taking a bait or lure.

These two examples indicate that individuals within a population of fish may be naturally more or less prone to taking a baited hook (more or less "catchable"), but not necessarily that individuals are actually learning to avoid hooks.

To consider cases of learned hook avoidance, we need to turn to controlled experiments. I am unaware of such experiments being carried out in Australia, but overseas studies on trout and pike in ponds have shown that fish are statistically less likely to be hooked again after only one catch-and-release. This effect is especially true for fish caught on lures (they are less likely to

Released fish may avoid re-hooking for some time.

take a lure a second time) although it is not as clear-cut for baited hooks. The interpretation is that an individual fish may learn from a once-off experience to avoid a foreign object like a lure, but is less likely to become overly cautious of a bait hiding a hook.

Do fish learn to avoid hooks? Yes … and maybe.

With the upsurge in catch-and-release fishing facilities in Australia, I would imagine operators have observations or data on hook shyness caused by single or multiple capture. The barramundi pond fisheries would be excellent sources of this kind of information.

I guess it's not hard to understand that hook and line fishing among a naive population of fish would catch the most susceptible fish first; the fish which, for genetic reasons, are not as wary as others, or which don't notice anything odd about a morsel of dead food with a hook in it. But, how does one explain the phenomenon sometimes observed in recreational fishing whereby a "new" method of fishing produces great results for a while, but then peters out, even though there is no way that the method has made any real dent in fish numbers? One particular example I'm thinking of was the phenomenon of high speed spinning for small pelagics from the rocks that took the fishing world by storm in the late 1970s. Rods, reels and lures all evolved rapidly to cater to the booming market and for a while it seemed that everyone involved with fishing had hitched up to the bandwagon. Then, as quickly as it all began, the foam settled and the boom was over. Why? The story I was told was that catch rates had fallen off dramatically and that it was no longer fun to thrash the water for hours for minimal result. Perhaps this was another case of quickly fishing down the catchable fish, but, somehow, I doubt it.

Finally, a story that is not about hook avoidance, but is about changing fish behaviour through fishing. If you visit Narooma on the New South Wales south

A black marlin caught with a circle hook, about to be released. This type of hook nearly always lodges in the corner of the mouth, causing no internal injury.

coast, you will notice a series of rock walls inside the mouth of beautiful Wagonga Inlet. These walls were built decades ago with a single purpose: to trap schools of travelling Australian salmon entering the inlet every year like clockwork. Sure enough, in the first year the salmon that entered the inlet were duly trapped, and the foundations of a reliable fishery had apparently been laid. However, the following year the schools of salmon seen approaching the inlet did not turn left, but proceeded on up the coast, apparently never coming into the port again in any numbers. I suppose this extraordinary event might also be explained by the initial culling of naive fish from the population but, again, this seems to me to be an unsatisfactory explanation. Even though we know a lot about the world of fish, there are still many phenomena continuing to defy explanation. That's what keeps us curious, and makes fishing so unpredictable. And that's not such a bad state of affairs.

Thoughts on Shark Finning

In the past, debates about fisheries usually dwelt on the question of sustainability. Consideration of the fate of individual fish was never really high on the agenda but, inevitably, that is certainly changing. Tuna long-lining has generally been conducted a long way from shore, so any environmentally undesirable aspects have tended to be ignored. Fortunately, long-lining does not kill very many non-fish species, such as whales, dolphins or turtles, so for many years it has been pursued largely in the absence of scrutiny. However, a big issue regarding long-lining has recently galvanised not just the conservationists but also the general public – the practice of finning of sharks caught as a by-product of the method.

Like many other exotic natural products, shark fins bring a high price on Asian markets, and the crews of long-liners are generally allowed to keep shark fins as a form of bonus. Apart from being seen as a waste of a whole fish, the primary problem with this is that the fins are, more often than not, removed from live sharks, which are then dumped back into the ocean (87 per cent of long-line caught sharks are alive at the vessel when the line is retrieved). As an example, the Hawaiian long-line fishery has expanded rapidly in recent years, and fins of sharks have become an important by-product for the crew. There are 110 active vessels in this fishery, and,

The Oceanic Blue Shark: the most commonly finned species in the world.

using observers as well as mandatory logbooks, the US authorities consider that their statistics on shark finning are reasonably accurate. During the period 1991 to 1999, an annual average of 80,000 to 100,000 sharks had been caught in this fishery. However, during this same period, the percentage of sharks finned has climbed from 10 per cent to more than 60 per cent, or in other words, over 60,000 sharks are now killed annually in this fishery for their fins alone. These numbers have caused great consternation, but things are never as simple as they seem. Over 95 per cent of sharks caught and finned are blue sharks, the most prolific shark of the open ocean, and a species that appears to be the exception to the rule of shark biology. Blue sharks grow quickly and mature early, and all indications are that their stocks in the Pacific are in good shape. So the issue of finning boils down to an ethical one, primarily revolving around the perception of cruelty. If the whole shark were utilised, or perhaps even if the shark was humanely killed before finning, the outrage would probably be far less intense. But should the fishery even be able to waste a carcass to retain the valuable fins?

There is a large market for dried shark fins. Are fisheries based on "luxury" items acceptable, even if sustainable?

We tend to automatically think it is acceptable to catch any fish (or invertebrate) commercially, provided that this is sustainable, and that the catch is used primarily for human consumption, so in this light, shark finning is seen as wasteful, if not abhorrent. But there are many other fisheries targeting a species not for the muscle (flesh, fillets) but for other parts, such as the eggs or roe. A classic example is the caviar fishery, which takes gravid female sturgeon for their hugely valuable eggs. And there is also one example closer to home: the mullet fishery of eastern Australia, which makes more money from selling the roe of pre-spawning fish than from the carcasses, which are often sold as trap bait. There are many other uses that are reasons for targeting fish commercially

that have very little, or nothing, to do with direct human consumption. Taking large quantities of small fish for fish meal is one example (used as fertiliser or stock feed additives), as is taking fish for bait (commercial and recreational), for feeding to captive tuna and salmon, or, of course, for pet food. And when you think about it, what about the world's tuna long-line fisheries themselves? Most of the tuna caught in this way end up in restaurants as small morsels of sashimi or sushi, mainly in Japan but increasingly in other countries as well. Expensive, gourmet morsels to be sure, but hardly feeding the ravenous hordes.

Shark finning is seen by many as totally unacceptable, but by others as a legitimate use of a sustainable resource.

Returning to the question of shark finning, many anglers would consider finning to be a purely commercial issue – but is it? There are some parallels with recreational fishing that should not be ignored. Sharks are a legitimate target species for anglers, in fact, I know many people who are particularly passionate about their shark fishing. But like the practice of finning, there is an issue over the fate of the shark once caught. Fortunately, these days more and more sharks are being filleted and eaten. Victorians have known for decades that shark flesh is delicious, but it's taken a bit longer for the message to be picked up in other states. Most of the sharks and rays are finding their way onto anglers' tables, but there is still a residual of sharks that are caught and weighed but not eaten. It can be argued, perhaps without a lot of conviction, that such a use of a natural resource is legitimate since the fishery still contributes to local economies, and numbers of sharks caught are very small.

The ethics of fishing are not easily defined. The issue of shark finning may seem clear-cut at first but, as with all things, there are always two sides to every story.

Index

Amphibious 127-129
Anchovy 183, *183*, 184
Anemones 29
Attacks 45, 154-156, 164, *164*, 165, *165*, 170

Bacteria 112, 113
Barramundi, *Lates calcarifer*, 81, 82, 174, *174*
Bass 118, 174, 181
  Red bass 31
Billfish,
  Istiophoridae, 13, 61, 64-65, 78, 79, *79*, 80, 82, 87, 90, 91, 92, 109, 111, 115, *115*, 116, 117, 123, 131, 141, 150, 176
Biodiversity 53-56, 191
Blennies 56, 118
Blue-green algae 31
Bonefish 57, *57*, 58
Bream 43, 113, 180, 62

Some fishes, such as the common lionfish, Pterois volitans, *warn would-be predators not to touch. Their spines can inject a powerful venom.*

Broadbill swordfish, *Xiphias gladius*, 12, 13, *13*, 14, *14*, 40, 45, 58, 61, 64, 65, 78, 87, 163, *165*
Bullrouts 26
Bundaberg disease 113

Camouflage 56, 71, 120
Cardinal fish, *Cheilodipterus macrodon*, 141, *141*
Carps 56, 108, 123
  Mahseer 56
Catch and release 43, 174, 175, 189-191, 192
Catfish 26, 27, 29, 56, 108, 118, 174
  Eel-tailed catfish, *Tandanus tandanus*, 106-107, *106*
  Fork-tailed catfish 141
  South America catfish, *Vandellia cirrhosa*, 108
  Striped catfish, *Plotosus lineatus*, 143,*143*
Candiru 56
Chinaman fish, *Symphorus nematophorus*, 31, 32, *32*
Cichlids 113, 141
Ciguatera 29-32
Clingfish, *Aspasmogasgter tasmaniensis*, 128, *128*, 129
Clown fish, *Amphiprion polymnus*, 140, *140*
Cobia, *Rachycentron canadus*, 61, 62, *62*, 138
Ccocoon 114, 149
Cod 56, 66, 67-68, 108, 110, 131, *131*, 139
  Butterfly cod 27, 29, 56
  Flowery cod, *Epinephalus fuscoguttatus*, 66, 67
  Orange spotted cod 67-68
  Potato cod, *Epinephalus tukula*, 67, 131, *131*
Coelacanth 74-75
Colombus, Christopher 136
Colour-vision 115-117

Cook, Captain James 48-49, 50
Cone shells 28, 29
Conservation 173, 174
Coral 26, 27, 55
Cosmopolitan fish 60-62
Crocodiles 29
Crossopterygian 75
Crustaceans 14, 17, 27, 55, 130, 166
Cuttlefish 134, 186

Deformities 43, 177, 178
DNA 55, 62, 83, 90, 92, 170
Dolphin fish , *Coryphaena hippurus*, 12- 15, *14*, *15*, 16, 17, 34, 36, 61, 82, 105, 109, 118, 119, 124
Dorsal fin 42, 43, 78, 79, 109, 137

Earbone, see otolith
Eel 56, 128
  Blue ribbon eel, *Rhinomuraena quaesita*, 55, *55*
  Moray 56, 114, 144, 123
  Southern conger eel 35, 56,
Eggs 34, 61, 81, 82, 83, 87, 89, 96, 121, 125, 126, 133, 134, 139, 140, 141, 143, 147, 166, 167, 180, 196

First aid 26
Fish sauce 183-185
Flathead 27, 59, 81, 82, 112, 180
  Dusky flathead, *Platycephalus fuscus*, 59, *59*, 125, *125*
  Tiger flathead, *Neoplatycephalus richardsoni*, 74, *74*
Flounder 56,
  Halibut (giant flounder) 44,
Flying fish, *Exocetus sp*, 46, *46*, 56, 102, *102*,

The endangered, and endearing Tasmanian red handfish, Brachionichthys politus. *The biodiversity of the seas is truly amazing.*

  103, *103*, 104, 119
  Fortescues 26, 27

Garfish (ballyhoo), *Dactyloptera spp*, 103, 139
  Gurnards 103, 105, *105*
Fortescues 26, 27

Gemfish 50, 144, 150, 181
Gill rakers 130, 131, *131*, 132, *132*
Glasseye, *Priacanthus blochii*, 132, *132*
Goatfish 108
Gobbies 56, 118
Goldfish 56, 110, 121, 123, *123*
Gropers 66-67, 68
  Blue groper 126, *126*
  Queensland groper 56, 67, *67*
Grunion 129, 139
Gurnards 27, 102, *204*

Hagfish 113
Herring 56, 130, 184
  Oxeye herring 50
  Tarpon herring 56
Humped batfish, *Platax batavianus*, 118, *118*

The amazingly coloured clown anglerfish, Antennarius maculatus. *Looking just like living sponge, it lies in wait for passing prey.*

Jackass morwong 50
Jellyfish *24*, 24-26, 27, 28, 29, 55, 148, 170
  Blue Bottle, *Physalia physalis*, 26, 28
  Box jellyfish, *Chiromex fleckeri*, 25, 26, 29, *29*
  Stockton Beach stinger 25
Jewfish 42, *42*, 43, *43*, 174, 177

Lampreys 163
Larvae 34, 61, 79, 83, 89, 96, 121, 131, 147, 148, 180, 181
Lionfish, *Pterois volitans*, 56, *200*
Long-lining 195
Louvar, *Luvarus imperialis*, 36, *36*, 37, *37*
Lungfish, *Neoceratodus forsteri*, 56, 123, 129, *129*

Mackerel 31, 42, 57, 109, 118, 131, 139
  Frigate mackerel, *Auxis thazard*, 61, 147
  Scaly mackerel, *Grammatorcynus bicarinatus*, 35, *35*
  Slimy mackerel , *Scomber australasicus*, 36, 47, 61, 110, 112, *112*, 114, *114*, 119
  Snake mackerel 62
  Spanish mackerel 45
Marlin 40, 56, 65, 76, *76*, 77, 81, 82, 83, 87, 90, 105, 114, 116, 118, 131, 136, 138, 150, 163, 177, 178, 179, *179*, 180
  Black marlin, *Makaira indica*, 34, 63, *63*, 73, 74, 77, *77*, 78, *78*, 79, 80, *80*, 81, *81*, 82, 98, *98*, 111, 115, 117, 118, 125, 138, *138*, 194, *194*, *203*
  Blue marlin, *Makaira nigricans*, 34, 43, 61, 64, 74, 78, 79, 81, 82, 90, 92, 110, 118, 131, 148, 165, 178, *178*
  Striped marlin, *Tetrapturus audax*, 38, 58, 79, 82, 90, *90*, 91, *91*, 92, 148,
  white marlin 82
Molluscs, *Cephalopods*, 27, 55, 134, 186
Moonfish, see opah
Mudskippers 128-129
Mullet 62, 108, 113, 123, 149, 196
Mulloway, see jewfish

Nautiluses 133, *133*, 134, *134*, 135, *135*

Oarfish 62
Octopus 28, 133, 134, 135
  Blue-ringed octopus 28, 29
Oilfish 62
Opah , *Lampris guttatus*, 61, *61*, 62
Orange roughy, *Hoplostesis atlanticus*, 121, *121*, 144
Otoliths 13, 18, 19, 20, 80, 85, 123

Parasites 15, 16, 17, 22, 23, 62, 95, 107, 108, 131, *131*, 137, 178,
Parrot fish 149, *149*
Pectorial fin 20, 102-103
Pelagic 12, 13, 34, 61, 62, 87, 115, 118, 146, 147, 148, 182, 193

Pelicans 187-188, *188*
Perches 56
Pike 123, 174, 192
Pilchards 56
Pineapple fish, *Cleidopus gloriamaris*, 34, *34*
Pinkling, *Genypterus tigerinus*, 113, *113*, 114
Pipefish, *Doryhamphus jansii*, 56, 69, 71, 72
Banded pipefish 71
Cleaner pipefish 71, 72, *72*
Ghost pipefish, *Solenostomos paradoxus*, 69, 71, *71*
Pipehorses 69
Piranhas 106
Pony fish 33, *33*, 34, 178, 184, *184*
Porcupine fish 56, *56*
Puffer fish 29, 56, 107, 118

Rabbit fish, *Siganus fuscescens*, 27, *27*
Rays 55, 124, 138, 197
Banjo ray 50
Manta ray 62, 138
Sting rays 28, *28*, 29
Red spot disease, *see* Bundaburg disease
Remora 136-138, *136*, *137*, *138*, 165, 166, *205*
Ribbon fish 62

Sailfish, *Istiophorus platypterus*, 50, 58, 61, 74, 78, 79, 82, 87, 92, *92*, 93, *93*, 94, *94*, 95, *95*, 96, *96*, 98, 109, 110, 116, 118, 131, 141
Salamander 127
Salmon 44, 56, 108, 123, 139, 174, 197,
Australian salmon, *Arripis trutta*, 49, *49*, 50, 194
Pacific salmon 107, 122
Threadfin salmon, *Polynemus plebeius*, 56, *56*
Sardines 46, 56
Scorpion fish *see* cod (butterfly cod)
Stone fish 56
Sea dragon 69, 71
Leafy sea-dragon, *Phycodurus eques*, 56, 69, 70, *70*
Weedy sea-dragon, *Pyllopteryx taeniolatus*, 69, *69*

The black marlin, Makaira indica, demonstrates its awesome power.

Sea snakes 28, 29, 147
Seahorse 56, 69, 71, 72, 139, *139*, 140
shorthead seahorse, *Hippocampus beviceps*, 72, *72*
Sea wasp 25
Sex 81, 82, 83, 124, 125, 126
Sexual dimorphism 81-83, 124, 126
Sharks 27, 29, 42, 55, 62, 65, 77, 124, 138, 152-171, 195-197
Basking shark 158, 166, 170
Blue shark 2, 158, 195, 196
Cookie-cutter shark, *Isistius brasiliensis*, 62, 163, *163*, 164-165, 170
Dwarf lantern shark 56
Goblin shark, *Mitsukurina owstoni*, 37, *37*, 38
Great White shark, *Carcharadon carcharias*, 62, 65, 154, *154*, 155, 159, *159*
Grey nurse shark 123, 155, 158
Hammerhead shark 62, 155, 158, 160, *160*, 161, *161*, 162, *162*
Mako shark, *Isurus oxyrinchus*, 41, 62, 64, *64*, 65, 111, *111*, 157, *157*

The red gurnard, Chelidonichthys kumu *uses its large, patterned pectoral fins for display, not for flying!*

Mega-mouth shark, *Megachasm pelagios*, 38, 62, 158, 169-171, *169*, *170*, *171*
Porbeagle shark 65, 158
Shark finning 195-197
Shovel nose shark 50
Thresher shark, *Alopias vulpinus*, 62, 65, *65*, 158
Tiger shark, *Galeocerdo cuvieri*, 62, 108, 118, 155, *155*, 158, *158*
Whale shark 62, 144, 155, 156, *156*, 157, 158, 166, *166*, 167, *167*, 168, *168*, 170
Wobegong shark 155
Shellfish 22, 26
Silver drummer 50
Slime 112, 113, 114
Snakehead, *channa sp*, 127, *127*
Snapper, *Pagrus auratus*, 18, 19, *19*, 21, 22, *22*, 23, *23*, 48, *48*, 50, 56, 134, 175
Soldier fish 144, *144*, 151, *151*
Spawning 13, 18, 34, 80, 82, 91, 107, 122, 123, 125, 129, 139, 141, 144, 147, 151, 180, 196
Spearfish 78, 79, 82, 131
Sponges 55, 186
Squid, *Sepioteuthis lessoniana*, 14, 39, *39*, 130, 134, 135, 150, 164
Giant squid, *Architeuthis dux*, 39-41, *41*, 135
Starfish 55
Stargazer, *Kathetostoma laeve*, 50, 73, *73*
Stonefish 26, 27, 28, 29
Streamlined 109, 111, 114, 178
Sturgeons 56, 108, 122, 123
Sucker fish, see remora
Sunfish 56, 62
Surgeon fish 27, 38, *38*, 118

Tagging 44, 91, 92, 93, 94, 95, 96, 97, 98, 99, 111, 112, 116, 168, 174, 176, 189

Tailor, *Pomatomus saltator*, 59, 177, 180
Tetras 56
Toad fish 37, 123
Trevally 87, 120, 139
Trout 56, 123, 139, 174, 175, 178, 192
Coral trout, *Plectropomus leopardus*, 30-31, *30*, *31*, *32*, 118, 192
Rainbow trout 110, 124
Tuna 56, 57, 63, 64, 65, 77, 87, 90, 91, 109, 111, 114, 115, 118, 123 130, 139, 141, 150, 163, 166, 178, 197
Bigeye tuna 61, 82, 88, 89, *89*, 110, *110*, 115, 130, 168
Blue fin tuna 43, 46, 60, 61, 110, 130, 178, 180
Dog tooth tuna, *Gymnosarda unicolor*, 35, 88, *88*
Longtail tuna, *Thunnus tonggol*, 34, 130, *130*
Skipjack tuna, *Katsuwonus pelamis*, 61, 63, 82, 87, *87*, 88, 89, 118, 120, 123, 145, *145* 147, 180
Yellow fin tuna, *Thunnus albacares*, 34, 36, 43, 61, 82, 84, *84*, 85, *85*, 86, *86*, 88, 89, *89*, 97, *97*, 98, 110, *110*, 115, 118, 130, 142, 148, 168
Turretfish, *Tetrosomus concatenautus*, 54, *54*
Turtle 136, 146, *146*, 186, 195

Urchins 55, 186

Wahoo, *Acanthocybium solandri*, 34, 61, 60, *60*, 110, 118
Walking Fish 56
Weever fish 29
Whale 45, 138, 144, 163, 187, 195
Baleen whale 144
Sperm whale 39-40, 46

A remarkable adaptation. The adhesive plate of the sucker fish, Remora remora, *is actually a highly modified dorsal fin.*

Wrasses 56, 71, 81, 124

Yellowtail kingfish, *Seriola lalandi*, 15, 16, *16*, 17, 59, 99, *99*, 134, 139, 177, *177*

Zebra fish, *Girella zebra*, 118, 119, *119*

Picture Credits

Kelvin Aitken: 10-11, 24, 25, 26, 28, 29, 52-53, 56, 100-101, 105, 111,142, 146, 152-153, 154, 157, 159, 167, 172-173, 195, 196 top, 196 bottom.
John Ashley: 18, 20, 21, 22, 86 top & bottom, 160.
Australian Museum Nature Focus: 37 bottom, 73, 74, 75.
Glen Booth: 84, 182, 191 bottom, 197.
Peter Bristow: 76, 77, 81, 93, 94 top, 96, 138, 155, 176, 203.
British Museum of Natural History: 148,149.
Greg Edwards: 63.
Lance Ferris: 188.
Greg Finney: 36, 37 top, 115.
Ken Graham: 65.
Barry Hutchins: 169, 171.
Sid Kraul: 14 top.
Rudie Kuiter: 19, 34, 39, 41, 55, 58, 69, 70 top & bottom, 71, 72 top & bottom, 99, 101 insert, 108, 116, 118, 120, 126 top & bottom, 128, 131, 132, 133, 134, 135, 139, 140, 141, 143, 144, 145 bottom, 149, 151, 161. 200, 201, 202, 204, 205.
John Matthews: 177.
Mark Mikkelson: 14 bottom.
Museum of Victoria: 41.
Damon Olsen: 78.
Brad Page: 168.
Julian Pepperell: 12, 15, 23, 33, 35, 45, 64, 79, 80, 82, 83, 87, 89, 92, 94 bottom, 95, 98, 103, 110, 117, 124 left & right, 125, 130, 137, 145 top, 147, 148, 150, 156, 158, 162, 164 top & bottom, 173 insert, 174, 175, 178, 180, 181, 185, 186, 187 top & bottom, 189, 190, 191 top, 192, 193, 194.
Random House Picture Library: 17.
Lester Rohrlach: 42, 43, 44, 166
Tim Simpson: 85.
Don Tuma: 13, 32, 38, 60, 62, 66, 68, 112, 129, 165.

All fish illustrations by Bernard Yau, except p50-51 (Craig Smith).

Every effort has been made to contact the copyright holder of all images.

Acknowledgements

A book such as this could never have become a reality without the help of a great many people. Sally Bird of Calidris Publishing Services was there at the beginning of the concept, and had the tenacity and enthusiasm to pursue the idea to completion. Glen Booth, editor of *Modern Fishing Magazine*, has gently edited my jottings for many years. He always allowed me considerable freedom in choice of topic and style, for which I am grateful.

At Random House, James Mills-Hicks fully supported the project from go to whoa, and remained calm in the face of impending deadlines.

The choice of images to illustrate each piece and chapter was very important, so I am especially grateful to Susan Rawkins and Lena Lowe for their innovative and creative work on the design of the book.

Special mention needs to be made of the marvellous images of photographers Rudie Kuiter and Kelvin Aitken, and illustrator Bernard Yau. Rudie and Kelvin have probably spent more time under the water than above it, and their photographs have certainly enhanced and illustrated many of the pieces to great effect.

Bernard Yau's illustrations of fish are a delight. He works with fresh subjects on his kitchen table, meticulously reproducing their finery in coloured pencil and ball-point pen, giving his work unique charm.

Thanks also to marine artist Craig Smith for his gorgeous illustration of a sailfish.

Many others encouraged me to keep going, including my sister, author Jenny Isaacs, father Geoffrey, Kate, Delse and good friend John Dunphy.

And of course, to the many hundreds of interested people who have written, telephoned, emailed or approached me with pictures, stories and questions about their observations of marine life, thank you for the opportunity to share some of the wonders of this world with you.